ROAD WORK

ROAD WORK

A New Highway Pricing and Investment Policy

Kenneth A. Small
Clifford Winston
Carol A. Evans

The Brookings Institution
Washington, D.C.

Library of Congress Cataloging-in-Publication data

Small, Kenneth A.
Road work: a new highway pricing and investment policy
Kenneth A. Small, Clifford Winston, and Carol A. Evans.
p. cm.
Includes index.
ISBN 0-8157-9470-3 (alk. paper)
1. Roads—United States—Finance. 2. Roads—Estimates—United States. I. Winston, Clifford, 1952- . II. Evans, Carol A., 1964- . III. Title.
HE355.S49 1989
388.1'14'0973—dc20 89-33528
CIP

9 8 7 6 5 4 3 2 1

The paper used in this publication meets the minimum requirements of American National Standard for Information Sciences—Permanence of Paper for Printed Library Materials, ANSI Z39.48-1984

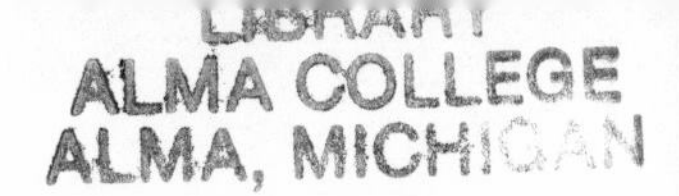

THE BROOKINGS INSTITUTION

The Brookings Institution is an independent organization devoted to nonpartisan research, education, and publication in economics, government, foreign policy, and the social sciences generally. Its principal purposes are to aid in the development of sound public policies and to promote public understanding of issues of national importance.

The Institution was founded on December 8, 1927, to merge the activities of the Institute for Government Research, founded in 1916, the Institute of Economics, founded in 1922, and the Robert Brookings Graduate School of Economics and Government, founded in 1924.

The Board of Trustees is responsible for the general administration of the Institution, while the immediate direction of the policies, program, and staff is vested in the President, assisted by an advisory committee of the officers and staff. The by-laws of the Institution state: "It is the function of the Trustees to make possible the conduct of scientific research, and publication, under the most favorable conditions, and to safeguard the independence of the research staff in the pursuit of their studies and in the publication of the results of such studies. It is not a part of their function to determine, control, or influence the conduct of particular investigations or the conclusions reached."

The President bears final responsibility for the decision to publish a manuscript as a Brookings book. In reaching his judgment on the competence, accuracy, and objectivity of each study, the President is advised by the director of the appropriate research program and weighs the views of a panel of expert outside readers who report to him in confidence on the quality of the work. Publication of a work signifies that it is deemed a competent treatment worthy of public consideration but does not imply endorsement of conclusions or recommendations.

The Institution maintains its position of neutrality on issues of public policy in order to safeguard the intellectual freedom of the staff. Hence interpretations or conclusions on Brookings publications should be understood to be solely those of the authors and should not be attributed to the Institution, to its trustees, officers, or other staff members, or to the organizations that support its research.

Foreword

AMERICA'S INTERSTATE HIGHWAY SYSTEM is deteriorating, and traffic congestion in most urban centers is worsening. Because of the many strong and conflicting interests, policy discussions about the road system are also in gridlock. The only consensus that seems to have emerged is that public spending must be increased.

In this study, Kenneth A. Small, Clifford Winston, and Carol A. Evans propose a comprehensive highway policy to meet the goals of efficiency, equity, and financial soundness. Their policy is based on two economic principles: efficient pricing to regulate demand for highway services and efficient investment to minimize the total public and private cost of providing them. Policy recommendations include a set of pavement-wear taxes for heavy trucks, a set of congestion taxes for all vehicles, and a program of optimal investments in road durability. Their proposals should be especially attractive to policymakers because they can be implemented with current technology, offer little threat to the major interest groups, and in the long run will reduce the strain on state and local governments' highway budgets.

Kenneth A. Small is a professor of economics at the University of California, Irvine; Clifford Winston is a senior fellow in the Brookings Economic Studies program; and Carol A. Evans is a former research assistant in the Brookings Economic Studies program. The authors are grateful to a long list of colleagues, especially at Brookings, for their helpful comments, and acknowledge the valuable perspectives provided by staff members of the U.S. Department of Transportation and the Florida Department of Transportation and by William Paterson of the World Bank. They would especially like to thank José A. Gomez-Ibañez, Herbert Mohring, and Philip A. Viton for their thoughtful and constructive reviews of the manuscript. Research assistance was provided by Heping He.

Brenda B. Szittya edited the manuscript, Leslie F. Siddeley and Anna M. Nekoranec verified its factual content, and David Rossetti provided staff assistance. The study was funded in part by a grant from the State Farm Insurance Company.

The views expressed in this book are those of the authors and should not be ascribed to those persons or organizations whose assistance is acknowledged or to the trustees, officers, or other staff members of the Brookings Institution.

BRUCE K. MACLAURY
President

April 1989
Washington, D.C.

Contents

Tables

Figures

ROAD WORK

CHAPTER ONE

Introduction

AMERICA'S ROAD SYSTEM is one of the world's most impressive public investments. Encompassing nearly 4 million miles, the system of federal, state, county, and local roads handles roughly 2 trillion vehicle-miles of travel each year. Its current value approaches one-half trillion dollars.[1]

Massive public spending is required each year for construction and upkeep of the road system. As shown in table 1-1, annual disbursements doubled from 1975 through 1985; by 1987, the nation's annual road bill was about $66 billion.[2] The table also reveals a disturbing trend. In 1975, road user taxes and tolls covered only 65 percent of total disbursements; ten years later the gap was even wider, despite a large infusion of tax revenues from the 1982 Surface Transportation Assistance Act. The responsibility for closing the gap falls to state and local governments, whose resources have been seriously drained. In 1985, as the table shows, local governments alone spent some $6.5 billion of their general revenues, $3.5 billion of their property taxes, and $4 billion in other revenues to help cover the shortfall in user taxes and tolls.

Furthermore, the highway sector's appetite for public funds shows every sign of continued growth. Forecasts of capital needs for the remainder of this century call for $38–$44 billion annually (1985 dollars), well above current capital outlays.[3] And as the last row of the table shows, merely maintaining existing roads accounts for a growing share of disbursements.

Despite its extent and cost, the nation's road system is deficient. According to the Federal Highway Administration, about 14 percent of nonlocal roads were in poor condition in 1985 and another 39 percent

1. National Council on Public Works Improvement, *Fragile Foundations: A Report on America's Public Works* (Government Printing Office, February 1988).

2. Motor Vehicle Manufacturers' Association, *MVMA Motor Vehicle Facts and Figures '87* (Detroit: MVMA, 1987), p. 83.

3. *The Status of the Nation's Highways: Conditions and Performance*, Committee Print, House Committee on Public Works and Transportation, 100 Cong. 1 sess. (GPO, 1987), page 76.

Table 1-1. *Road Revenues and Disbursements, 1975, 1980, 1985*
Billions of dollars

Item	*1975*	*1980*	*1985*
Revenues			
Road user taxes and tolls	18.65	25.55	35.60
Road user taxes	17.39	23.89	33.41
Federal	5.87	9.52	11.57
State	11.33	14.11	21.31
Local	0.19	0.26	0.53
Bridge and ferry tolls	1.25	1.65	2.19
Other revenues	10.68	16.07	25.43
General funds	4.57	8.40	9.88
Federal	1.46	2.45	2.04
State	0.57	1.37	1.39
Local	2.55	4.58	6.45
Property taxes	1.65	2.34	3.47
Federal	...	...	...
State	...	...	...
Local	1.65	2.34	3.47
Miscellaneous and other imposts	2.22	3.27	6.20
Federal	0.79	1.07	1.40
State	0.81	1.28	2.95
Local	0.63	0.92	1.85
Bond proceeds	2.24	2.06	6.08
Federal	...	...	..
State	1.44	1.15	3.92
Local	0.80	0.91	2.16
Total revenues	29.33	41.62	61.03
Disbursements			
Capital outlays	14.38	20.30	26.57
Maintenance	7.29	11.41	16.59
Administration, police, and other	7.01	10.02	14.30
Total disbursements	28.68	41.73	57.46
Road user taxes and tolls as percentage of disbursements	65.0	61.2	62.0
Maintenance as percentage of disbursements	25.4	27.3	28.9

Source: Federal Highway Administration, *Highway Statistics* (Department of Transportation, various years), tables HF-1, HF-2. Breakdown of other revenue sources by level of government based on percentages in table HF-10.

were in only fair condition.[4] Poorly maintained roads may cause an estimated $30 billion in vehicle damage and additional fuel expenditures each year.[5] Road congestion is also severe and on the rise, particularly in major metropolitan centers. Currently, nearly half of all urban interstates and one-third of other main urban arterial highways have highway capacity use rates above 70 percent, a threshold indicator of congestion.[6] The New York Port Authority estimates that commuting delays to and from Manhattan have doubled during the 1980s, and the Washington, D.C., Council of Governments estimates that rush-hour speeds on the Capital Beltway have fallen 15–50 percent during the same time.[7] It has even been argued that inadequate public infrastructure spending, which has contributed to the rise in travel time and expense, is partly responsible for the decline in the nation's productivity during the 1980s.[8] With traffic expected to continue its rapid growth, the United States can anticipate continued strain on its pavements, bridges, and road capacities.

Historical Overview

Federal involvement in the nation's roads can be traced to the 1916 Rural Post Roads Act, which authorized federal grants to pay for up to half the costs of constructing rural roads used to deliver the mail.[9] The large share of road development costs borne by the federal government has engendered debate about how users should be charged and how

4. Ibid., pp. 33, 39.

5. This estimate, made by The Road Information Program (TRIP), a Washington, D.C., organization of highway builders and users, is reported in Richard Corrigan, "Rebuilding Our Highways," *AAA World,* vol. 5 (July–August 1985), pp. 4–5.

6. Federal Highway Administration, *Highway Statistics, 1986* (Department of Transportation, 1987), table HM-61.

7. Richard Levine, "Car Madness in Manhattan: Cure Sought," *New York Times,* October 11, 1987, p. E-6; John Lancaster, "Beltway Rush-Hour Commuters Find the Going Getting Slower: COG Panel Acts to Spur Wilson Bridge Work," *Washington Post,* April 21, 1988, pp. D1–2.

8. Alan S. Blinder, "Are Crumbling Highways Giving Productivity a Flat?" *Business Week* (August 29, 1988), p. 16.

9. Complete histories of the evolution of the highway system and highway pricing and finance are given by Philip H. Burch, Jr., *Highway Revenue and Expenditure Policy in the United States* (Rutgers University Press, 1962); and Congressional Budget Office, *New Directions for the Nation's Public Works* (GPO, September 1988). The material in this section is adapted from these sources.

costs should be allocated among different classes of users, primarily between trucks and autos.

Initially, federal highway programs were financed entirely from general revenues. In 1932, the federal government imposed a tax on gasoline fuel, the revenue from which was formally earmarked for highway programs when the Highway Trust Fund was created in 1956. In the interval, the states, which had led the way in imposing gasoline taxes and which had also introduced vehicle registration fees, took a closer look at the allocation of pavement-wear costs between trucks and autos. Following World War II, eleven states adopted "third-structure" taxes, which attempted to assess heavy vehicles according to their total weight and distance traveled. During the 1950s and 1960s, the third-structure tax generated more interest and debate than any form of highway taxation. What sparked the debate was not the yield of the tax—less than 1 percent of all highway revenue—but its potential impact on the trucking and bus industries.

In 1961, the federal government released its influential Highway Cost Allocation Study, which concluded that certain heavy trucks were not bearing their full financial responsibility in some states, but were paying more than their share in others, especially those imposing third-structure taxes.[10] This and succeeding cost-allocation studies have been contested by truckers and other highway interests, and there is still no consensus on what constitutes an appropriate road user charge. To be sure, the federal government's thinking has evolved from a fixation with fuel taxes to an interest in taxes that more accurately reflect a vehicle's contribution to pavement damage. But the evolution has been slow, and, as we shall argue, inappropriate user charges have been partly responsible for current problems with the road system.

During the 1970s and 1980s, federal attention has focused on rehabilitating the highways. Beginning in 1974, federal funds were made available for so-called 3R projects (restoration, resurfacing, or rehabilitation) on federal-aid highways. In 1981, the program was amended to include reconstruction. Revenues were to come primarily from funds generated by the Surface Transportation Assistance Act of 1982, which raised the tax on fuel from 4 cents to 9 cents a gallon, revised and increased the tire tax, raised truck sales taxes, and substantially raised the use tax on

10. U.S. Bureau of Public Roads, *Final Report of the Highway Cost Allocation Study*, 87 Cong. 1 sess., House Doc. 54 (GPO, 1961).

heavy vehicles. Truckers mounted such heavy opposition to the increased use tax that the Deficit Reduction Act of 1984 reduced it and instead raised the tax on diesel fuel still further, to 15 cents a gallon. Despite this infusion of revenues, current projections still indicate an annual shortfall of funds throughout this century.[11]

Although highway user charges and finance have evoked considerable public debate, a closely related issue, highway design, has generated little disagreement. The federal government took an early role in contributing to highway research and planning. Section 1 of the Hayden-Cartwright Act in 1934 stipulated that up to 1.5 percent of a state's federal aid allotment could be used for these purposes. One important result was the road test conducted by the American Association of State Highway Officials (AASHO) between 1958 and 1960, the findings of which continue to be influential in the design guide for highway construction issued by the American Association of State Highway and Transportation Officials (AASHTO), the successor to AASHO. The design guide has evolved as the test's findings have been critically examined and extended, but AASHTO has never wavered in its commitment to the basic test results. Neither it nor the federal government has seriously explored the effects that changes in highway design would have upon the cost-allocation debate. We argue that the inattention to this issue (perhaps due to tremendous efforts required simply to build and maintain the system) has contributed to current problems with the road system.

Highway Interests

A public investment as important to the national economy as the highway infrastructure naturally generates strong and conflicting interests among users. Successful highway policy must take into account the diverse interests of the federal government, state and local governments, the railroad, trucking, and mass transit industries, the driving public, and highway contractors.

Federal, state, and local governments are under intense pressure to enlarge and improve the road system while keeping down taxes. This is

11. For further discussion, see Howard Gleckman, Tom Ichniowski, and Richard Hoppe, "The Crumbling of America: A New Crisis in Public Works," *Business Week* (December 1, 1986), pp. 62–63; *Fragile Foundations: A Report on America's Public Works*.

Table 1-2. *Federal and State Road User Taxes, 1986*
Billions of dollars

Tax	*Revenue*
Federal road user tax[a]	
Motor fuel tax	10.25
Truck and trailer tax	1.14
Tire tax	0.32
Other tax	0.53
Total	12.25
State road user tax	
Motor fuel tax	14.74
Registration fees	8.42
Other fees and miscellaneous receipts	4.58
Total	27.74

Source: *Highway Statistics, 1986,* tables FE-9, MF-1, MV-2.
a. These taxes comprise the payments into the Highway Trust Fund.

not easy. On the expenditure side, even earmarking revenues for highways does not fully insulate highway finance from short-term political and budgetary considerations, especially at the federal level. On the revenue side, the dominant form of road user tax is the fuel tax, levied at both federal and state levels and accounting for more than 60 percent of all user taxes (table 1-2). The fuel tax has not always proven a reliable source of revenue; since 1973 receipts have fluctuated along with economic conditions and fuel prices. Except for the sharp increase in 1982, when the federal fuel tax rate went up 125 percent, the recent fuel-tax revenue trend has been generally downward in real terms (because of improved fuel economy and increased use of untaxed gasohol).[12] Yet despite the occasionally severe erosion of real revenues, states have found it politically difficult to raise gasoline taxes.

Meanwhile, road users and other interest groups are demanding better roads and shifts in the financial burden. The American Automobile Association and the railroad industry want trucks to pay a higher share to offset the damage they inflict. The trucking industry counters that most road damage is caused by weather and inadequate design. The mass transit industry tries to maintain its share of the Highway Trust Fund. Finally, road contractors want a major highway program that is free of federal appropriation delays.

12. For data on the tax increase, see Bureau of the Census, *Statistical Abstract of the United States, 1985* (Department of Commerce, 1984), p. 594.

Making Policy for the Future

Current highway policy, shackled by these conflicting interests and by past policy, appears to lack clear objectives and rationale. The only consensus that has emerged is that public infrastructure spending must be increased. But even assuming that large sums of money were available, the nation should not invest more in its highways without looking carefully at how such investments are made and how the road system is priced. Despite extensive discussion and testimony, a number of questions remain open with no satisfactory mechanism for resolution. Are user taxes set efficiently and can they improve the condition of the U.S. road system? Are the taxes fair? Is the mix of expenditures on road maintenance, capacity, and durability appropriate? Could and should these expenditures be fully financed by efficient user taxes? Can a more stable source of revenue be provided?

With federal user taxes scheduled to expire in 1993 and congressional decisions on highway programs and user taxes required by 1991, with congestion receiving almost daily media attention, and with deteriorating roads awaiting funds for repair, it is time to consider alternatives to current policy. This book proposes a comprehensive highway policy to meet the goals of efficiency, equity, and financial stability. It can be implemented using current technology, and it offers little or no threat to the major highway interest groups. The policy entails a set of pavement-wear taxes for heavy trucks, a program of optimal investments in road durability, and a set of congestion taxes for all vehicles. The pavement-wear taxes would depend on axle loads rather than on vehicle weight or, as federal policy has historically emphasized, on fuel consumed; the congestion taxes would depend on a vehicle's contribution to congestion. Optimal investments in road durability would produce mostly thicker roads, especially in the cases of heavily traveled interstates and other principal arterials. We estimate that the pavement-wear taxes and optimal investments in road durability could generate $8 billion (1982) in annual net benefits to society, largely accrued in reduced depreciation that would lead to lower future maintenance expenditures. Congestion charges could produce additional net benefits of some $5 billion annually, mostly in the form of time savings to motorists. The combination of new revenues and reduced long-run costs of highway provision would eventually make it possible to reduce greatly the nation's dependence upon fuel taxes, registration fees, and general revenues of state and local governments to finance its roads.

The policy's import, however, goes far beyond fiscal savings and technicalities of highway construction and finance. A nation's standard of living is inextricably tied to its transportation infrastructure. By significantly enhancing it our policy will enable America's citizens to obtain far greater fulfillment from their working and nonworking activities.

CHAPTER TWO

Road Pricing and Investment

ROAD USER CHARGES and optimal investment, though often treated separately by policy analysts, are facets of the same problem: both are aimed at minimizing the total costs of building, maintaining, and using a road system. Although investment pertains to the initial design and construction of a road and user charges pertain to ongoing user and maintenance activities, the two are interdependent.

The best way to economize on maintaining and using an existing road is to apply a user charge equal to the actual cost each user imposes on society through his effect on the road's condition and on the speed that other users can travel. Such a charge, known as the marginal-cost user charge, ensures that the independent decisions by users reflect the interests of all. Marginal cost is defined as the *change* in the total social cost of travel on existing roads, including costs of road maintenance and costs incurred by all its users, brought about by adding one vehicle of a particular type and weight at a particular place and time. If road users are required to incur this entire amount themselves, they will use the highway (at that time and place) only if the value to them of doing so exceeds the amount society must pay, in aggregate, to accommodate them. Low-priority uses are thereby deterred and high-priority uses are accommodated with fewer deleterious effects from crowding or pavement deterioration.

The social cost of road use depends also upon the design of the road. The better the road, the smaller the cost of use and the associated marginal-cost user charge. Optimal investment policy focuses on this trade-off: capital is initially invested in the design and construction of the highway up to the point where any further investment would cost more than the resulting savings. This trade-off ties investment policy to user charges, or pricing policy, and points to another advantage of marginal-cost user charges: the resulting revenues provide a tangible signal to public officials as to whether additional investments to provide more or better service are likely to be worthwhile.

Road investment has two dimensions: capacity and durability. Capacity is needed to accommodate the vehicle flow without excessive congestion and is typically increased by adding lanes, making them wider, improving shoulders, improving intersections, building a median, or adding overpasses. Durability—long-term pavement serviceability—is needed to accommodate a cumulative flow of heavy vehicles without excessive pavement damage and the accompanying extra costs to both the public treasury and highway users. Durability is typically increased by making the pavement thicker, improving the base beneath the pavement, improving drainage, using better materials, using steel reinforcing rods, improving joints between sections of concrete, or tightening construction tolerances.

Because both types of road investment are expensive, optimal investment will involve some scarcity, both of capacity and of durability. Pricing is a natural economic response to scarcity, and the two types of scarcity lead to two types of user charges. Charges associated with scarce capacity, which causes congestion, should reflect a vehicle's contribution to congestion. Charges associated with scarce durability, which causes road wear (that is, pavement deterioration), should reflect a vehicle's contribution to this wear.

Engineers and policymakers concerned with highway finance have focused on road wear, largely ignoring the important role of congestion. Economists, in contrast, have paid a great deal of attention to congestion, much less to road wear, and almost none to the two combined. We deal with the two combined in order to study the financial and political viability of user charges, but we give the greatest attention to road wear because it has been relatively neglected in economic analyses.

Precise theory for our study requires mathematics, but the concepts and intuition are not difficult. We provide a nontechnical exposition in the text and a mathematical exposition in the appendix.

Optimal Pricing: Marginal-Cost User Charges

The two primary considerations in pricing the use of existing roads are road-wear costs and congestion costs. Other social costs such as noise, air pollution, and dependence on foreign energy supplies may also be important, but we do not address them.

Road-Wear Costs

Road-wear costs to be taken into account in pricing road use include maintenance costs incurred by the agency responsible for the road, and user costs incurred by operators of both cars and trucks, including repair, vehicle depreciation, fuel, and their value of time. To maintain consistency with standard terminology in the literature on road wear, we restrict the term user costs to those that occur under free-flowing traffic conditions—that is, they exclude congestion costs.

The dominant element in maintenance cost is the periodic overlay that is required when a pavement becomes worn. To a first approximation, the pavement is designed to withstand a certain number of passages of axles of a standard weight and configuration before requiring an overlay.[1] The standard is a single axle of 18,000 pounds; the damaging power of an axle with some other load or configuration is then defined in terms of the number of "equivalent standard axle loads" (esals) causing the same damage. A heavy concrete pavement typical of interstates on major trucking routes will withstand the passage of about 9 million such standard axles or their equivalent before requiring an overlay.

One component of marginal cost, then, is the shortening of the period between overlays. We call this the marginal maintenance cost of road wear, denoted MC_m. Because other maintenance expenditures on paved roads are small by comparison, we ignore them.

Two technological facts are crucial to understanding road wear. First, the equivalence factor for an axle rises *very* steeply with its load—roughly as its third power. (As explained in the appendix, it was previously thought to rise as the fourth power, and the relationship has become known as the "fourth-power law.") Thus, for example, the rear axle of a typical thirteen-ton van causes over 1,000 times as much structural damage as that of a car; if illegally loaded to nineteen tons, it would cause at least three times more damage. For all practical purposes, structural damage to roads is caused by trucks and buses, not by cars.

Second, it is the weight per axle that matters, not total vehicle weight. A 50,000-pound two-axle dump truck causes more road wear than a huge twin-trailer rig spreading 100,000 pounds over seven axles. A major goal of pricing policy should be to reduce the heaviest *axle weights*. None of

1. Federal Highway Administration, *Final Report on the Federal Highway Cost Allocation Study* (Department of Transportation, 1982), p. IV-42.

the charges now used accomplishes this, nor do most versions of the weight-distance taxes currently under study by congressional mandate.[2] Some charges are downright perverse. Charging by the axle, as do many turnpikes and the state of Ohio, encourages carrying heavy loads on too few axles; so do the federal excise tax on truck tires and the fuel-tax surcharge on multi-axle trucks in Kentucky and Virginia.[3]

Worn, rough pavement generates not only maintenance costs, but costs incurred by road users through vehicle damage and, to a lesser extent, reduced speeds. The average user cost over an entire overlay cycle is unaffected by the cycle's duration; but when discounted to the present in order to account properly for compound interest, these user costs are higher if heavy traffic shortens the cycle. We call this the marginal user cost of road wear, or MC_u.

Added together, MC_m and MC_u constitute the marginal cost of road wear, MC_w. A user charge based on axle weight should approximate this measure. Because the life of a pavement depends strongly on the pavement design, MC_w varies widely from road to road. We explore the implications of adopting a user charge based on MC_w in the following two chapters.

Congestion Costs

Congestion is caused by vehicle traffic measured in "passenger car equivalents" (pces) per hour. Each vehicle's pce is determined by the amount of road space it effectively takes up, including the space between vehicles required for safety, compared with that of an average car. A truck or bus typically has two to five passenger car equivalents, depend-

2. However, there has been some interest in the so-called "Turner proposal" advocated by Francis C. Turner, former Federal Highway Administrator, which would adjust weight limits so as to permit higher gross weights while reducing individual axle weights. See Stephen R. Godwin and others, "Increasing Trucking Productivity within the Constraints of Highway and Bridge Design," *Transportation Quarterly*, vol. 41 (April 1987), pp. 133–50. Although the Turner proposal does not address pricing, it is certainly a step toward rationalizing the system of weight limits.

3. Gary R. Allen, "Highway User Fees: Are These Old Taxes Still Good Taxes?" in American Association of State Highway and Transportation Officials, *Understanding the Highway Finance Evolution/Revolution* (Washington, D.C.: AASHTO, 1987), pp. 27–34 (especially p. 32).

ing on the type of road and terrain. Congestion is mostly caused by cars, though lane blockage due to truck accidents is also a problem.[4]

Engineers and other researchers have spent considerable effort quantifying the relationship between traffic volume and speed, in order to measure the extra delay caused by adding one passenger car equivalent to the traffic stream. Combined with estimates of vehicle operating costs and the values that people place on their time, that measure of extra delay allows us to measure the increase in costs to all other users due to this additional car. We call this increase in costs the marginal cost of congestion, denoted MC_c.

Congestion costs vary not only from road to road, but also by time of day. This crucial feature further complicates the implementation of user charges, and has led to considerable study, new technology, and some experimentation.[5] It is possible to estimate a congestion charge based on an average of costs of many different roads and at many different times, but most economists believe that, to be useful, congestion pricing should be more narrowly targeted on congested roads at peak hours only.[6]

Optimal Investment

Roads come in all shapes and sizes. A road can be a single narrow lane with occasional turnouts, or a twelve-lane expressway. It can be made of dirt, gravel, thin asphalt, or a foot of Portland cement concrete on top of crushed-gravel base and cement-stabilized subsoil. The only reason to choose more expensive designs is to reduce subsequent road-

4. That congestion is mostly caused by cars was recently verified for California in a study by Cambridge Systematics, Inc., as reported by Douglas Shuit, "Caltrans to Seek Rules for Trucks on Freeways," *Los Angeles Times,* January 21, 1989.

5. For an excellent review, see Steven A. Morrison, "A Survey of Road Pricing," *Transportation Research,* vol. 20A (March 1986), pp. 87–97.

6. For an estimation of an average congestion charge, see David M. Newbery, "Road User Charges in Britain," *Economic Journal,* vol. 98 (Conference 1988), pp. 161–76. William Vickrey represents the mainstream of economists in words he wrote a quarter of a century ago: "there can be no efficient solution to the urban traffic problem that does not include provision for charges on automobile use that are differentiated according to time of day." See William S. Vickrey "General and Specific Financing of Urban Services," in Howard G. Schaller, ed., *Public Expenditure Decisions in the Urban Community* (Washington, D.C.: Resources for the Future, 1962), pp. 62–90.

wear and congestion costs. As noted earlier, road-wear costs are strongly affected by a road's durability, and congestion costs by its capacity.

The road's thickness, *D*, is the primary determinant of its ability to accommodate numerous heavy axle loads. Making the road thicker is expensive, but it saves road-wear maintenance and user costs. At the margin, a highway designer wanting to minimize total life-cycle cost would add thickness (or other measures to improve durability) until the incremental cost of additional durability equaled the incremental saving in maintenance and user costs. That incremental saving is just another facet of the marginal cost of road wear. Hence a user charge for road wear can be viewed as a charge for providing the durability needed to handle heavy axles.

In our empirical work in chapter 3 and its appendix, we compute optimal durability by simply minimizing the sum of annualized maintenance cost, user costs (where included), and capital cost. Maintenance and user costs are assumed to go on forever, but are discounted to the date of construction. There are many components to the capital cost of a road, but the only one that affects these calculations is the incremental cost of adding additional thickness, which is determined from the portion of highway contracts for the paving material itself and its installation. The explicit manner in which costs depend upon variables such as road thickness and pavement life is specified in the appendix to this chapter.

The road's width, *W*, measured in number of lanes, is the primary determinant of its capacity, that is, the number of cars per hour that can travel on it before extreme congestion sets in. Making the road wider is expensive, but it lowers congestion costs. At the margin, it makes sense from an efficiency standpoint to add width (or other capacity-improving measures) until the incremental cost of additional capacity equals the incremental congestion-cost saving that it brings about. That incremental saving is just another facet of the marginal cost of congestion, so the user charge for congestion can be viewed as a charge for providing the capacity needed to keep congestion to a tolerable (that is, efficient) level. We compute optimal capacity under selected conditions in chapter 6.

Formulas for Road-Wear Costs

To see how these pricing and investment principles could work in practice, we need explicit empirical models of components of highway technology. In the case of road-wear costs, for example, we model such things as pavement deterioration rates, maintenance strategies, and the variation of user costs with pavement quality. In this and the following section, we provide the intuition underlying these models, which are derived formally in the appendix.

The easiest way to begin is with what we term the "naive" marginal maintenance cost of road wear. Suppose N is the number of standard axle passages that the right-hand lane can withstand before requiring an overlay, and λ is the fraction of one-directional axle passages that travel in that lane. For practical reasons, all lanes of a multilane roadway are nearly always built with the same thickness and resurfaced at the same time;[7] hence when cumulative one-directional traffic loadings on the entire roadway reach N/λ standard axle passages, the right-hand lane will have incurred N passages and the entire roadway is resurfaced. Let C be the cost of this overlay, including the cost of temporary disruptions to traffic. Then the average maintenance cost per axle passage is

$$MC_m^0 = \frac{C}{N/\lambda} = \lambda C/N.$$

The most recent federal highway cost allocation study estimated MC_m^0 (in 1981 prices) to be 9 cents per esal-mile for rural interstate highways, 66 cents for urban arterials, and 80 cents for urban local streets.[8] Thus, for example, a truck equivalent to 1.5 standard axles traveling 100 miles on a rural interstate would accrue 150 esal-miles with an associated cost of 150 times 9 cents, or \$13.50. (We give these figures as examples, though we believe that the resurfacing cost used to obtain these estimates is overstated.)

Now consider a correspondingly naive marginal user cost of road wear. Total user costs rise and fall with changes in pavement quality. Suppose the road is always resurfaced at some trigger value of pavement quality. Then user costs vary cyclically between the same limits regard-

7. José A. Gomez-Ibañez and Mary M. O'Keeffe, *The Benefits from Improved Investment Rules: A Case Study of the Interstate Highway System,* Report DOT/OST P-34/86/030 (Department of Transportation, July 1985), p. C-10. At least two states, California and Florida, are now experimenting with construction and maintenance practices that leave the right-hand lane more durable than other lanes. We are unable to provide estimates of how much extra cost this practice entails, hence cannot tell how much our results would be altered if the practice became widespread.

8. FHWA, *Cost Allocation Study,* p. E-25, table 3.

less of the overlay cycle's duration, so that increased traffic affects the periodicity, but not the average value, of user costs. Hence the naive marginal user cost is zero:

$$MC_u^0 = 0.$$

Although these calculations completely ignore compound interest, they are less naive than one might think. David Newbery has shown that under certain circumstances, MC_u can be ignored and MC_m^0 is precisely the user charge which, if applied to a large number of roads distributed evenly among pavements of all ages, equals the marginal cost of a trip covering all those roads.[9] Hence if we advocated a pricing policy only, and wanted a single user charge to apply to roads with pavements of all ages, this would be a reasonable charge to choose.

However, we also advocate an investment policy, and we wish to exploit the theoretical link between the revenues from marginal cost pricing and the expenditures needed to undertake optimal investment. Because the investment decision can be made only at the beginning of a pavement's life, we need an estimate of the marginal cost that is properly discounted (that is, adjusted for compound interest) so as to apply beginning when the pavement is new. Adjusting for the fact that a dollar spent several years from now is worth less than a dollar spent today lowers MC_m and raises MC_u.

It lowers MC_m (below its naive value of MC_m^0) because the discounted cost of having to resurface the road is less than the value C used in MC_m^0. As a result,

$$MC_m = \alpha(MC_m^0),$$

where α is a fraction between zero and one that depends on the interest rate and the time between overlays. The formula for α is given in the appendix of this chapter; at a real interest rate of 6 percent and a pavement life of fifteen years, $\alpha = 0.94$.

Compound interest raises MC_u (above its naive value of zero) because the average user cost, when discounted to the time the road is built, depends on traffic in a positive way. Although the simple average of user cost over the entire cycle is constant, the highest user costs occur late in the cycle (just before the trigger value is reached) and hence are less important when discounted. The longer the cycle, the less the discounted

9. David M. Newbery, "Road Damage Externalities and Road User Charges," *Econometrica*, vol. 56 (March 1988), pp. 295–316.

value of those higher costs. Hence by shortening the cycle, an additional axle load raises this average discounted user cost. Put more simply, a heavy-axle passage today hastens the time when users start to hit potholes, thereby raising the importance of those users' future unhappiness. A more precise argument, and the resulting formula for MC_u, is given in the appendix.

We ignore user cost in the calculations of chapter 3 because the parameters needed to calculate it are so imperfectly known that it would make all our results less reliable. However, in the appendix to chapter 3 we recalculate our basic results using the best estimate of user cost we can provide. It appears that accounting for the variation of user costs with pavement quality would actually strengthen our main conclusions.

Road maintenance itself is a complex subject treated in a number of recent studies.[10] Of the numerous possible maintenance strategies, here we adopt the simple but common rule of thumb used in most studies of road wear: resurface whenever pavement quality reaches a specific value.[11] We take that value to be 2.5 as measured by the "Pavement Serviceability Index," a quality measure developed in connection with an influential road test.[12] There is no reason to think that our results are

10. A. Bhandari and others, "Technical Options for Road Maintenance in Developing Countries and Their Economic Consequences," *Transportation Research Record,* no. 1128 (1987); Gomez-Ibañez and O'Keeffe, *Improved Investment Rules;* Wayne S. Balta and Michael J. Markow, *Demand Responsive Approach to Highway Maintenance and Rehabilitation,* vol. 2: *Optimal Investment Policies for Maintenance and Rehabilitation of Highway Pavements,* Report DOT/OST/P-34/87/054 (Dept. of Transportation, June 1985); D. W. Potter and W. R. Hudson, "Optimisation of Highway Maintenance Using the Highway Design Model," *Australian Road Research,* vol. 11 (March 1981), pp. 3-16; David R. Luhr, B. Frank McCullough, and Adrian Pelzner, "Development of an Improved Pavement Management System," *Fifth International Conference on the Structural Design of Asphalt Pavements,* vol. 1: *Proceedings* (University of Michigan and Delft University of Technology, 1982), pp. 553–63; Organization for Economic Cooperation and Development, *Pavement Management Systems* (Paris: OECD, 1987).

11. As examples, see FHWA, *Cost Allocation Study;* and Newbery, "Road User Charges in Britain."

12. This index is defined by a formula incorporating objective measures of longitudinal profile, cracking, and patching. The formula was designed to approximate an earlier measure, known as the "pavement serviceability rating," which was an average of subjective ratings by a panel of experts. These subjective ratings were on a scale of zero to five, with 0–1 meaning "very poor" and 4–5 "very good." See Highway Research Board, *The AASHO Road Test: Report 5, Pavement Research,* Special Report 61E (Washington, D.C.: National Research Council, 1962), pp. 12–15, 143, 291–306; and American Association of State Highway and Transportation Officials, *AASHTO Guide for Design of Pavement Structures* (Washington, D.C.: AASHTO, 1986), pp. I-7 through I-9.

significantly biased by failing to derive this maintenance rule from first principles.

Pavement Technology

To apply the methodology we have outlined for analyzing pricing and investment, we must know how pavement quality is affected by axle weights and road thickness, as well as by aging and weathering.

Heavy Loads and Road Wear

The American Association of State Highway and Transportation Officials (AASHTO) has been a worldwide leader in sponsoring and interpreting research on heavy loads and road wear, although many other organizations and nations have made important contributions.[13] AASHTO's design guide is extensively used, especially in the United States.[14]

One empirical test stands out for its wide influence and provides the basic equations used in the AASHTO design guide. It was carried out in Illinois between 1958 and 1960 on specially constructed test loops, in which pavement design and truck loads were independently varied to see how each affected pavement wear. It is known as the AASHO road test, named for the American Association of State Highway Officials, the predecessor organization of AASHTO.[15]

In an earlier paper, two of us examined the data analysis in the original AASHO study to estimate the design equations.[16] We concluded that the statistical methods were flawed, and we reestimated a key relationship using modern econometric tools. The reestimation resulted in similar but better-fitting equations that predict considerably shorter lifetimes for the heaviest concrete pavements—a prediction corroborated by

13. See, for example, Organization for Economic Cooperation and Development, *Full-Scale Pavement Tests,* Road Transport Research (Paris: OECD, 1985).

14. *AASHTO Guide for Design of Pavement Structures.*

15. *AASHO Road Test.*

16. Kenneth A. Small and Clifford Winston, "Optimal Highway Durability," *American Economic Review,* vol. 78 (June 1988), pp. 560–69.

independent evidence. These equations and some of the corroborating evidence are described in the appendix.

The key equation is fitted separately for two classes of pavement, rigid and flexible, a categorization that follows virtually universal engineering practice. Rigid pavements are made of Portland cement concrete, a very hard light-colored surface, commonly called concrete, that tends to crack rather than deform when subjected to stress. Flexible pavements are of asphaltic concrete, commonly called asphalt or blacktop, a material with some compressibility especially in warm weather. Both pavements can be damaged not only directly by stress from above, but also by motion of the underlying soil, which is affected by stresses on the pavement itself and by moisture and temperature conditions underneath.

Like the AASHO researchers, we measured rigid-pavement strength directly as the thickness (in inches) of the concrete slab, and flexible pavement strength as a weighted combination of the thicknesses of the pavement itself, the base on which it is placed, and the subbase underlying that (see appendix). This weighted combination is called the structural number, but for convenience we refer to it as road thickness. We also followed the AASHO researchers by using their seasonal weights for flexible pavements to account for seasonal variation in soil strength; they found no strong seasonal pattern to rigid-pavement wear, so no such adjustment is needed when analyzing rigid pavements.

Our equations, like AASHO's, are fitted from data in a climate with strong seasonal variation and severe winter conditions. The guides used by highway designers contain methods to adjust the AASHO equations for different climate and soil conditions, but these are too complex for us to incorporate into a compact model of pricing and investment. Hence we base our calculations not so much upon road thickness itself as upon differences between current and optimal lifetime, which should be more transferable from one climate to another. By using this strategy, we believe that our model accurately evaluates the changes in investment criteria that we suggest; we would not claim, however, that the road thicknesses themselves resulting from our calculations represent optimal design in every locality.

Aging and Weathering

The road wear caused by heavy axles is exacerbated by time and weather, especially in severe climates. Wet subsoils are less resistant to the force transmitted through a pavement from a heavy axle. Cracks in the road surface are widened by frost, allowing water to damage lower parts of the highway structure, thereby lessening its ability to withstand future loads. Asphalt loses its flexibility over many years, becoming brittle and more likely to crack under loads.

Some analysts, especially members of the trucking industry, suggest that a substantial fraction of the total maintenance cost of a highway is attributable to age and weather. They then go on to claim that the extent of these costs invalidates the argument for user taxes that are steeply graduated by axle weight. In fact, the opposite is true. All evidence indicates that these effects are mainly interactive—it is weathering of the pavement in conjunction with heavy axles that causes the problem. (We review this evidence in the appendix.) Hence aging and weathering leave a pavement more vulnerable to damage by heavy loads, thereby raising the extra maintenance cost caused by those loads.

Extensive research by the World Bank has recently quantified this effect and verified its interactive nature. From data on asphalt pavements in Brazil, William Paterson estimates that pavement roughness (defined according to a precise international standard that correlates well with AASHO's measure of pavement quality) increases with age at a constant rate, in addition to the increase caused by traffic. Since roughness of a new pavement is nearly zero on this scale, this age-related increase would be very small in the absence of heavy loads; but any roughness caused by traffic tends to become magnified by age. The rate of roughness increase, m, was estimated at 2.3 percent a year for Brazil. Cruder estimates for other climates range from about 1 percent a year in dry warm climates (Tunisia) to 7 percent a year for moist freezing climates (Colorado), though the upper range is very uncertain.[17] We use an intermediate value of 4 percent.

17. William D. O. Paterson, *Road Deterioration and Maintenance Effects: Models for Planning and Management,* Highway Design and Maintenance Standards Series (Johns Hopkins University Press for the World Bank, 1987), pp. 306–16. Paterson suggests (p. 315) that even higher values may apply in wet climates with extreme freezing; but his only quantitative evidence, purporting to show that roughness in an untrafficked lane in the AASHO road test increased at an average rate of 23 percent a year, is flatly contradicted by the AASHO researchers themselves, who conclude that "no significant serviceability loss was found in the flexible pavements . . . that were not subjected to traffic over the 2-year period of the Road Test" (*AASHO Road Test,* p. 123). Paterson does not explain which of the AASHO data he used, and he may have included one very thin section that the AASHO researchers say was "inadvertently damaged" (pp. 122–23).

We verify formally in the appendix that the marginal maintenance cost of road wear is higher, not lower, when aging and weathering are present. On asphalt pavement with a fifteen-year life, weathering increases this marginal cost by 1 percent in Tunisia, by 39 percent in Colorado, and 14 percent at our intermediate value of m.

For rigid pavements, there is little or no evidence for any aging effects and plenty of evidence for very long pavement lives in the absence of heavy axle loads. We support these claims in the appendix. Hence we assume that m is zero on this important portion of the nation's heavy-duty arterial system.

Conclusion

This completes our description of the basic theory behind our proposed comprehensive highway policy. Full details are in the appendix.

One part of the policy addresses pavement wear and entails two significant changes in current practice. First, to foster better use of existing highways' load-carrying capabilities, we propose heavy-vehicle user charges that rise steeply with axle weights. Second, to further reduce total social costs, we propose an investment policy that increases road durability where doing so would sufficiently reduce future maintenance costs. We shall see in the next chapter that these two changes together not only harm truckers far less than would user charges alone, but also greatly improve the fiscal balance of public agencies.

The other part of the proposed policy addresses congestion. To better use existing highway capacity and lessen the need for costly additional capacity, we propose congestion charges that discourage travel at the most congested times and places. The implications of this proposal, which is likely to raise more political opposition than the proposals on road-wear charges and investment, are discussed in chapter 5.

Our focus so far has been the supply side of highway economics. We have presented an analysis of costs, which is all that is needed to derive the principles for pricing and investment. To predict the results of adopting these principles, we need to model the demand side. In the case of road-wear charges, we need to answer the question: how do truckers and shippers respond to changed pricing conditions? To that we turn next.

Appendix

This appendix provides a technical presentation of the model of pricing and investment used in subsequent chapters, as well as supporting detail for certain key assumptions.

We begin by adapting the standard economic model of congestion pricing and investment in road capacity, developed over many years through the work of Dupuit, Pigou, Knight, Mohring, and others, to include durability.[18] This modification yields a pricing rule with two components: a congestion charge related to scarce capacity and a heavy-vehicle charge related to scarce durability.

Suppose a one-mile, one-directional stretch of highway is used y days a year by vehicles in type-weight classes labeled $i = 1, \ldots, I$, during distinct hours of the day labeled $h = 1, \ldots, H$. Let $q = \{q_{ih}\}$ denote the hourly flow volumes, and let $P_{ih}(q)$ be the inverse demand functions. P_{ih} is an inclusive price—that is, it includes user-incurred time and money costs plus user charges.

Each vehicle in class i contributes as much to congestion as ϕ_i cars, and as much to road wear as μ_i single axles weighing 18,000 pounds: it is said to have ϕ_i passenger car equivalents and μ_i equivalent standard axle loads. We follow common practice by ignoring variation in these equivalence factors with terrain, climate, and highway design. We therefore define two aggregate traffic variables that affect congestion and road wear, respectively: hourly traffic *volume*, $V_h = \Sigma_i \phi_i q_{ih}$, and annual traffic *loadings*, $Q = y \Sigma_h \Sigma_i \mu_i q_{ih}$.

Let W and D measure the road's capacity and durability, respectively. They are continuously variable, but for convenience we measure W as pavement width in units of twelve feet, the width of a standard lane, and D as pavement thickness in inches (or a weighted combination of various component thicknesses). We denote the present values of highway maintenance and capital costs for our one-mile highway segment as $M(Q,W,D)$ and $K(W,D)$, respectively.

Let u_{ih} be the annualized present value of the per-mile cost to user i at hour h. It has two components: user cost, v_i, which would be incurred

18. For work of earlier researchers, see the reviews by Clifford Winston, "Conceptual Developments in the Economics of Transportation: An Interpretive Survey," *Journal of Economic Literature*, vol. 23 (March 1985), pp. 57–94; and Morrison, "A Survey of Road Pricing."

under free-flow conditions and hence depends upon how quickly the road deteriorates; and congestion cost, c_{ih}, which is the additional time and money cost caused by congestion delays. Road deterioration depends on Q and D, whereas congestion during hour h depends on V_h and W; hence the cost to users is

$$u_{ih}(V_h,W,Q,D) = v_i(Q,D) + c_{ih}(V_h,W).$$

(Congestion could also depend on traffic volumes at earlier times, because of queuing; this would complicate the equations but not the concept behind them.)

It is useful to define an aggregate user cost equal to the present value of all user costs:

$$U(Q,D) = \frac{y}{r} \sum_h \sum_i q_{ih} v_i(Q,D),$$

where r is the real interest rate. We also define an aggregate congestion cost for each hour, h,

$$c_h(V_h,W) = \sum_i q_{ih}\, c_{ih}(V_h,W);$$

and an annual aggregate congestion cost,

$$c(V_1, \ldots, V_H,W) = y \sum_h c_h(V_h,W).$$

(Later in this appendix, we show how to calculate U if we know how the user cost in a given year depends on pavement quality during that year. In chapter 6, we will suggest a formula for c in the case where there is just one peak period.)

In accordance with the road pricing and investment literature, the net benefits to be maximized are equal to the annual consumer surplus from travel less all annualized costs.[19] Formally, we solve the following problem:

$$\text{(2-1)} \qquad \underset{q,W,D}{\text{Max}}\; y \sum_h \sum_i \int_0^{q_{ih}} P_{ih}(q')dq'_{ih} - y \sum_h \sum_i q_{ih} u_{ih}(V_h,W,Q,D) - rM(Q,W,D) - rK(W,D).$$

This yields the following first-order conditions:

$$\text{(2-2a)} \qquad P_{ih} - u_{ih} = \phi_i(\partial c_h/\partial V_h) + \mu_i[r\partial(M+U)/\partial Q],$$

$$i = 1, \ldots, I, h = 1, \ldots, H,$$

19. The same first-order conditions can be derived more laboriously by maximizing,

$$r\partial(M+K)/\partial W = -\partial c/\partial W, \tag{2-2b}$$

$$\partial K/\partial D = -\partial(M+U)/\partial D \tag{2-2c}$$

(hence a present value such as U or M is annualized by multiplying it by r, the real interest rate).

Equation 2-2a, the pricing rule, shows that a user charge is needed to make up the difference between the costs u_{ih} already borne by the user and the inclusive price P_{ih} required for demand to be at its optimal level. This user charge is given by the right-hand side of the equation; it consists of a congestion charge (proportioned to ϕ_i) plus a charge for road wear (proportioned to μ_i), the latter reflecting costs borne both by the highway authority and by other motorists. (Note that this formula assumes the charge for road wear cannot be varied over the life of the pavement). Equation 2-2b, the optimal capacity rule, equates the extra cost of building and maintaining a wider road to the incremental benefit of reduced congestion. Equation 2-2c, the optimal durability rule, equates the extra capital cost of building a thicker road to the incremental benefit of reduced maintenance and user cost.

We now examine how M, K, and U depend on traffic loadings and road thickness. This will allow us to derive specific formulas for those components of equations 2-2a and 2-2c that relate to road wear. Equation 2-2b and the other part of equation 2-2a, relating to congestion, are deferred until chapter 6.

Pavement Technology

We provide here an analysis of road wear adapted from previously published work by Kenneth Small and Clifford Winston.[20]

We denote by N the number of esals that, under conditions of negligible aging, causes pavement quality π to decline to the critical level π_f, at which time an overlay is required. This is simply a technological property of the pavement and is directly measured in accelerated road tests such as the AASHO test described in the text.

---

subject to an economywide budget constraint, a Bergson-Samuelson social welfare function for which an increment of income to any person has the same marginal social value.

20. Small and Winston, "Optimal Highway Durability," pp. 562–65.

The researchers who analyzed the AASHO road test specified a nonlinear equation relating a precisely defined measure of pavement quality, π, to the cumulative number of applications n of an axle weight L_1 (measured in thousands of pounds) and type L_2 (defined as $L_2 = 1$ for single axles, 2 for tandem; a tandem axle is two axles close together). For flexible pavements, where seasonal differences in pavement vulnerability are large, n is a seasonally weighted number of applications, the weights having been developed from a separate analysis. The equation is

$$\pi = \pi_0 - (\pi_0 - \pi_f)(n/N)^b, \quad \text{(2-3)}$$

where π_0 is initial pavement quality and π_f is a predetermined "terminal" pavement quality at which the pavement is considered to be worn out. For π_f, AASHO used 1.5, but we use 2.5, a value more typically used as the trigger value in the decision to resurface. The quantities N and b depend parametrically on L_1, L_2, and D as described below. By setting $n = N$, we see that N is just the number of axle passages that will cause the pavement to wear out (that is, reach quality π_f).

For each of the 548 experimental pavement sections, parameters b and N were estimated by AASHO using ordinary least squares applied to equation 2-3 in the form

$$\log\{(\pi_0 - \pi)/(\pi_0 - \pi_f)\} = [-b\log(N)] + [b]\log(n), \quad \text{(2-3a)}$$

with the square brackets indicating the intercept and slope of the regression, respectively. However, bounds were placed on permissible values of b.

The AASHO researchers then specified equations to explain how b and N varied across pavement sections. Their equation for N was

$$N = A_0(D + 1)^{A_1}(L_1 + L_2)^{-A_2}(L_2)^{A_3}. \quad \text{(2-4)}$$

For rigid pavements, D is just the pavement thickness in inches. For flexible pavements, D is a linear combination of pavement, base, and subbase thicknesses with coefficients 0.44, 0.14, and 0.11; it is known as structural number.[21] The coefficients A_i were estimated separately for rigid and flexible pavements, the unit of observation being one of 264

21. The coefficients defining structural number were simultaneously estimated along with other parameters in a complex multistage procedure; *AASHO Road Test,* pp. 36–40. For AASHO estimates of these equations, see *Road Test,* pp. 40, 152.

rigid or 284 flexible experimental pavement sections, and using the previously estimated values of N as dependent variables.

This cumbersome procedure has unknown statistical properties and is almost certainly not statistically efficient. Small and Winston therefore reanalyzed the data using a simple, standard technique to estimate equation 2-4 only. Its left-hand side is observed directly, permitting consistent and efficient estimation of that equation (after taking logarithms). However, many pavement sections outlasted the experiment, so the observed dependent variable is $\min\{N, n_{max}\}$ where n_{max} is the number of axle passes applied during the experiment; this necessitated using the censored regression (Tobit) estimation method with censoring at an upper limit of n_{max}. A more complex analysis, using all the data and estimating b as well, has recently been undertaken for rigid pavement, with no important changes in the results used here.[22]

Our estimates of equation 2-4 are shown along with AASHO's in table 2-1. Recall that the dependent variables are slightly different, so the two sets of estimates are not strictly comparable. Nevertheless, some important differences are apparent. First, the bottom row shows that our estimation procedure fits the relevant data points better that AASHO's. Second, our estimates show a somewhat less steep relationship between pavement life and axle load (coefficient A_2)—closer to a third-power law than to the fourth-power law conventionally used to approximate the AASHO findings. Our estimates also show a somewhat less steep relationship between pavement life and pavement thickness (coefficient A_1) than AASHO's.

The most important difference, however, is seen by calculating N, the pavement life in esals. This is done for our estimates simply by substituting $L_1 = 18$ and $L_2 = 1$ into equation 2-4. Doing so reveals that our estimates imply far shorter pavement lifetimes for thick pavements than do AASHO's—as much as 65 percent shorter (9.3 million versus 26.6 million esals) for the standard ten-inch rigid slab used on most interstates.

There is considerable evidence corroborating our finding that the AASHO equations overstate lifetimes of thick pavements. An early critique noted that AASHO's estimated curves for thick flexible pave-

22. Kenneth A. Small and Feng Zhang, "A Reanalysis of the AASHO Road Test Data: Rigid Pavements," paper presented to the Econometric Society, New York, December 1988.

Table 2-1. *Estimates of Equation 2-4*

Variable	Coefficient	Rigid pavements: Small-Winston[b]	Rigid pavements: AASHO[c]	Flexible pavements[a]: Small-Winston[b]	Flexible pavements[a]: AASHO[c]
Constant	$\ln A_0$	13.505 (0.307)	13.47	12.062 (0.237)	13.65
$\ln(D + 1)$	A_1	5.041 (0.329)	7.35	7.761 (0.245)	9.36
$-\ln(L_1 + L_2)$	A_2	3.241 (0.260)	4.62	3.652 (0.147)	4.79
$\ln(L_2)$	A_3	2.270 (0.242)	3.28	3.238 (0.189)	4.33
Summary statistic					
Number of observations	...	264	...	284	...
Number of censored observations[d]	...	191	...	45	...
Standard error of regression[e]	...	0.367	...	0.651	...
Standard error of prediction[f]	...	0.370	0.515	0.629	0.673

Source: Kenneth A. Small and Clifford Winston, "Optimal Highway Durability," *American Economic Review*, vol. 78 (June 1988), p. 564, table 1.

a. Using seasonally weighted axle applications.

b. Estimated using the Tobit model of error structure. Dependent variable is the natural logarithm of the number of axle applications to pavement serviceability index of 2.5. Standard errors are given in parentheses.

c. From Highway Research Board, *The AASHO Road Text: Report 5, Pavement Research,* Special Report 61E (Washington, D.C.: National Research Council, 1962), pp. 40, 152. The intercept has been converted from base ten to natural logarithms. Dependent variable was the logarithm of an estimated parameter representing the number of axle application to pavement serviceability index of 1.5. Standard errors were not reported.

d. A censored observation is one for which only a lower bound on the dependent variable is observed because of the finite duration of the test.

e. Estimated standard deviation of the error term in the log form of equation 2-4, obtained as part of the Tobit maximum-likelihood parameter estimation.

f. Root mean squared deviation between predicted and observed natural logarithm of number of axle loads yielding $\pi_f = 2.5$, for those observations where $\pi_f = 2.5$ was reached (73 such observations for rigid pavements, 239 for flexible). For each of the "AASHO" columns, the AASHO estimate of all three equations, namely equations 2-3a and 2-4 and an equation explaining parameter b, was used in the calculation.

ments fit the data poorly.[23] Two studies of heavy-duty rigid pavements in Illinois, one of them involving sections of the road test itself (which were later incorporated into Interstate Route 80), found that the AASHO equations overpredicted actual pavement lives by factors of two to three.[24] (The same was not true of thinner rigid pavements, so the discrepancy cannot be explained by weather or bad traffic forecasts.)

23. Canadian Good Roads Association, *Report of the Observer Committee of the Canadian Good Roads Association on the AASHO Road Test* (Ottawa: CGRA, 1962), p. 130.

24. Robert P. Elliott, "Rehabilitated AASH(T)O Road Test: Analysis of Performance

It is nevertheless important to recognize that the estimated lifetimes for thick pavements, whether from our procedure or AASHO's, involve extrapolation well beyond the range of direct observation. The road test was run for just over a million axle passes, whereas a thick pavement typically lasts 6 million to 10 million passes of a standard axle. In order to assess the robustness of this extrapolation, and more generally to test the sensitivity of our results, we tried a variety of other specifications of equation 2-4, including redefining the dependent variable by choosing $\pi_f = 1.5$; replacing $(D + 1)$ by D and $(L_1 + L_2)$ by L_1; using a translog functional form; and segmenting the sample by pavement thickness. For rigid pavements the results were not affected much by any of these changes, adding to our confidence in the conclusions we draw. For flexible pavements, results varied considerably (perhaps because of the seasonal weighting scheme), so we regard them as less certain.

Some writers have stressed the independent effects of aging and weathering on pavement deterioration. Gomez-Ibañez and O'Keeffe, after reviewing conflicting claims in the literature, conclude that time and weather mainly aggravate the effects of traffic loadings, having little independent effect.[25] But Paterson argues that the independent effect is substantial, at least on asphalt pavements in moist freezing climates.[26] Paterson's conclusion is based on a model estimated from a sample of in-use Brazilian roads, then adjusted to fit data from other climates. The Brazilian data, from a humid warm climate, give an aging effect of only 2.3 percent a year, which implies a traffic-free pavement lifetime of something like fifty years. Paterson recognizes the extrapolation involved in assigning aging effects to other climates, especially the most severe ones (humid freezing), and cautions that his model may not apply there.

We know of no evidence of significant aging in rigid pavements, and there is considerable anecdotal evidence that even thin rigid pavements can last well over thirty years when not subjected to heavy loads. The Pasadena Freeway in southern California, originally built as the auto-

Data Reported in Illinois Physical Research Report 76," *Paving Forum* (Summer 1981), pp. 3–9; Bob H. Welch and others, "Pavement Management Study: Illinois Tollway Pavement Overlays," *Transportation Research Record,* no. 814 (1981), pp. 34–40.

25. Gomez-Ibañez and O'Keeffe, "Improved Investment Rules," pp. C-5 and C-6, especially note 11.

26. Paterson, *Road Deterioration and Maintenance Effects,* section 8.6, pp. 306–316.

only Arroyo Seco Parkway, lasted thirty-five years without resurfacing. At that time, its left-hand lanes of thin flexible pavement had finally weathered enough to require rehabilitation; its right-hand lanes, of 6.6–9-inch-thick Portland cement concrete, were still sound.[27] Parts of the Wilbur Cross Parkway in Connecticut, an 8-inch rigid pavement carrying no trucks, lasted for thirty-five years, and most of the Merritt Parkway in Connecticut was overlaid only after thirty years and largely because of damage by studded snow tires.[28] A well-known British text on highway design recommends that rigid pavements be built to last at least forty years.[29]

Maintenance Cost

We now derive formulas for the present value of resurfacing costs as a function of traffic loadings Q and the parameter N describing pavement durability.

First, we must determine T, the overlay interval, as a function of Q and N. Suppose loadings are applied to the entire one-directional roadway at annual rate Q, and a proportion λ of them occur in the most heavily traveled lane. In the absence of aging effects, the right-hand lane and hence the entire road would require resurfacing at intervals of T^0 years, where

$$T^0 = \frac{N}{\lambda Q}. \tag{2-5}$$

More generally, we follow Paterson and Newbery by assuming that pavement roughness grows linearly with cumulative esals and exponen-

27. J. A. Mathews and K. L. Baumeister, *Damage to Pavement Due to Axle Loads,* Report No. CA-DOT-TL-3158-1-76-64 (Sacramento: California Department of Transportation, 1976), p. 10.

28. W. Ronald Hudson and Stephen B. Seeds, "Examination of Pavement Deterioration in the Presence of Automobile Traffic," *Transportation Research Record,* no. 993 (1984), pp. 55–62.

29. David Croney, *The Design and Performance of Road Pavements* (London: Her Majesty's Stationery Office, for the U.K. Transport and Road Research Laboratory, 1977), p. 25.

tially with time.[30] When used with Paterson's linear transformation that relates his measure of roughness and our measure of pavement quality,[31] this assumption implies that pavement quality obeys the following modified version of equation 2-3:

$$\text{(2-3b)} \qquad \pi(t) = \pi_0 - (\pi_0 - \pi_f)(\lambda Qt/N)e^{mt}.$$

T can be determined as a function of N and Q by setting $\pi(T) = \pi_f$ in equation 2-3b; T is then the solution of the following equation:

$$\text{(2-5a)} \qquad T = \frac{N}{\lambda Q} e^{-mT}.$$

Next, we show how M, the present value of all overlay costs, depends on T. Let $C(W)$ be the cost of an overlay. Then M is the present value of an infinite sequence of expenditures $C(W)$ every T years, beginning in year T. (We assume the overlay interval is constant.) Thus:

$$\text{(2-6)} \qquad M(Q,W,D) = \frac{C(W)}{(e^{rT} - 1)}.$$

The marginal maintenance cost can now be found by differentiating annualized maintenance cost, rM, with respect to annual traffic loadings, Q:

$$\text{(2-7)} \qquad MC_m = r\frac{\partial M}{\partial Q} = r\frac{\partial M}{\partial T}\frac{dT}{dQ} = -\frac{r^2 e^{rT} C(W)}{(e^{rT} - 1)^2} \cdot \frac{dT}{dQ}.$$

30. Paterson (*Road Deterioration and Maintenance Effects,* chap. 8) describes a long and complex sequence of analyses attempting to obtain satisfactory results from very recalcitrant data. We believe the "unconstrained aggregate level model" of his table 8.5 is preferable to the "component incremental model" of his table 8.3, which is not a reduced form (that is, it contains intermediate measures of surface distress that are themselves caused by traffic and aging). Newbery ("Road Damage Externalities," p. 307) makes the same choice in his equation 14 as we do. The preferred model, in our notation, is

$$R(t) = \left[R_0 + \frac{X(t)}{N}\right]e^{mt},$$

where $R(t)$ is roughness, $R_0 = R(0)$, and $X(t)$ is cumulative standard axle loads in the most heavily traveled lane at time t.

31. Paterson (*Road Deterioration and Maintenance Effects,* p. 43, equation 2.5) gives this relationship as $R = (9.2) - (2.2)\pi$. This implies that, for a pavement with $\pi_0 = 4.2$, $R_0 = 0$, and the equation in the previous footnote reduces to our equation 2-3b when $X(t) = \lambda Qt$.

Consider first the case of no aging, that is, $m = 0$. T is given by equation 2-5, from which $dT/dQ = -\lambda T^2/N$ so that

$$MC_m = \frac{\alpha C}{N/\lambda} = \alpha(MC_m^0), \tag{2-8}$$

where $\alpha = (rT)^2 e^{rT}/(e^{rT} - 1)^2$. We have shown elsewhere that $0 < \alpha \leq 1$.[32]

When m is not 0, we can find dT/dQ by differentiating both sides of equation 2-5a with respect to Q, and solving for

$$\frac{dT}{dQ} = -\frac{\lambda T^2}{N} \cdot \beta, \tag{2-9}$$

where

$$\beta = \frac{e^{mT}}{1 + mT}. \tag{2-9a}$$

In that case, the marginal cost is

$$MC_m = \alpha\beta(MC_m^0). \tag{2-9b}$$

Expanding e^{mT} in a Taylor Series shows that $\beta > 1$; hence, for a given pavement lifetime T, MC_m is higher than it would be in the absence of aging. For $T = 15$, simple calculation from equation 2-9a shows that $\beta = 1.01$ when $m = 0.01$, $\beta = 1.14$ when $m = 0.04$, and $\beta = 1.39$ when $m = 0.07$.

Capital Cost

Following several previous studies, we assume that construction cost is a linear function of width.[33] The portion that varies with width includes the cost of preparing the roadbed, which does not depend on pavement thickness, and the cost of the pavement itself, which is approximately

32. Small and Winston, "Optimal Highway Durability," p. 561, note 5.

33. John R. Meyer, John F. Kain, and Martin Wohl, *The Urban Transportation Problem* (Harvard University Press, 1965); Marvin Kraus, Herbert Mohring, and Thomas Pinfold, "The Welfare Costs of Nonoptimum Pricing and Investment Policies for Freeway Transportation," *American Economic Review*, vol. 66 (September 1976), pp. 532–47; Marvin Kraus, "Indivisibilities, Economies of Scale, and Optimal Subsidy Policy for Freeways," *Land Economics*, vol. 57 (February 1981), pp. 115–21.

proportional to the volume of paving material. A plausible specification is

(2-10) $$K(W,D) = k_0 + k_1 W + k_2 WD.$$

User Cost

In this section, we provide the detailed calculation of how average user cost varies with traffic loadings, still under the assumption that traffic loadings are applied at a uniform rate (growth in loadings is analyzed separately in the next section).

The reason average user cost depends on traffic loadings and pavement durability is that it is an annualized present value, not just an ordinary average. Let $V(t)$ be the traffic volume in year t and let $v(t)$ be the average user cost in year t. Because of pavement deterioration, $v(t)$ varies cyclically as shown in figure 2-1.[34] If $V(t)$ grows at a constant rate g—that is, $V(t) = V_0 e^{gt}$—then the present value of total user cost over the first cycle is

$$P_0 = \int_0^T V_0 e^{gt}\, v(t) e^{-rt}\, dt;$$

the present value over the second cycle is

$$\int_T^{2T} V_0 e^{gt} v(t - T) e^{-rt} dt = e^{-(r-g)T} P_0;$$

and so forth. (The cycle duration remains constant because of the rule that repaving is done when pavement quality reaches the predetermined level π_f.) The sum of all these, assuming $r > g$, is

(2-11) $$U = \frac{P_0}{1 - e^{-(r-g)T}} = \frac{V_0}{1 - e^{-aT}} \cdot \int_0^T v(t) e^{-at} dt,$$

where $a = r - g$.

34. Similar diagrams appear often in the literature on road maintenance: for example, D. W. Potter and W. R. Hudson, "Optimisation of Highway Maintenance Using the Highway Design Model," *Australian Road Research,* vol. 11 (March 1981), p. 9; Michael J. Markow and Wayne S. Balta, "Optimal Rehabilitation Frequencies for Highway Pavements," *Transportation Research Record,* no. 1035 (1985), p. 32; Newbery, "Road Damage Externalities," p. 301.

Figure 2-1. *Cyclical Variation in Average User Cost*

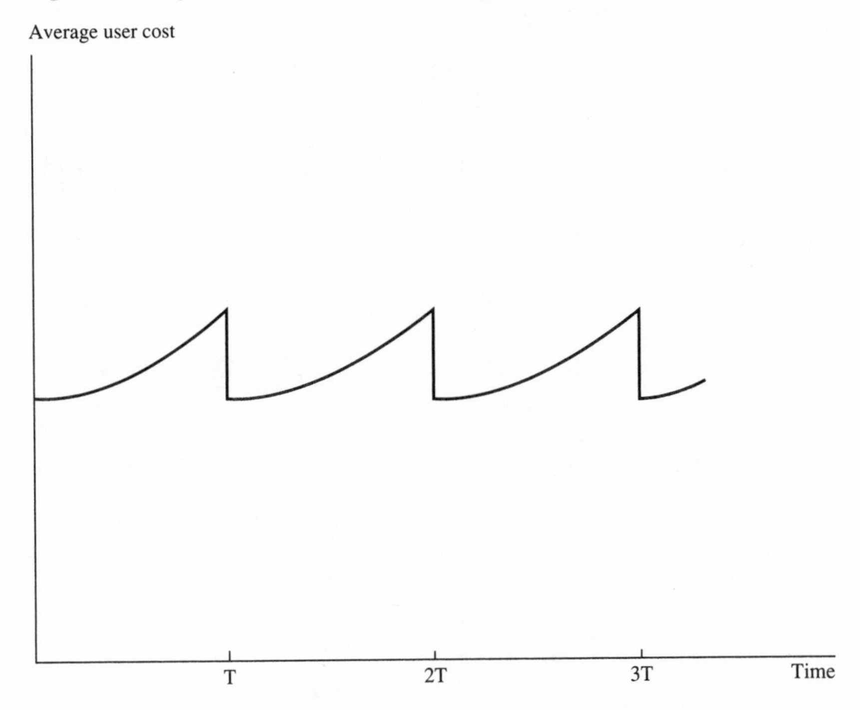

The shape and magnitude of $v(t)$ is quite uncertain, although the evidence suggests it is convex as shown in the figure. Our best estimate of its shape comes from an econometric study of rigid pavements that extends our earlier work on the AASHO road test data. That study estimated the parameter b in equation 2-3 to be 5.0, with standard error approximately 1.7.[35] Within the relevant range of pavement quality, average user cost seems to be approximately linear in $(\pi_0 - \pi)$, according to Newbery and the evidence amassed by Paterson.[36] Therefore

$$v(t) = v_0 + v_1 (t/T)^5, \tag{2-12}$$

where v_0 is the user cost on a new pavement, and $(v_0 + v_1)$ is the user

35. Small and Zhang, "A Reanalysis," table 1.

36. See Newbery, "Road Damage Externalities," p. 315, equation A8; Paterson, *Road Deterioration and Maintenance Effects,* section 2.1.2. This also uses the approximately linear relationship between Paterson's roughness measure and our serviceability index as given by Paterson, p. 43, equation 2.5.

cost on a pavement at quality π_f. Substituting equation 2-12 into the integral in equation 2-11 gives

$$(2\text{-}13) \qquad U = \frac{rV_0}{a}\left\{v_0 + v_1\left[\frac{120}{(aT)^5} - (e^{aT}-1)^{-1} \cdot \left(1 + \frac{5}{aT} + \frac{20}{(aT)^2} + \frac{60}{(aT)^3} + \frac{120}{(aT)^4}\right)\right]\right\},$$

from which

$$(2\text{-}14) \qquad \frac{dU}{dT} = \frac{rV_0v_1}{y}\left[-\frac{600}{y^5} + ye^y(e^y-1)^{-2}\left(1 + \frac{5}{y} + \frac{20}{y^2} + \frac{60}{y^3} + \frac{120}{y^4}\right) + (e^y-1)^{-1}\left(\frac{5}{y} + \frac{40}{y^2} + \frac{180}{y^3} + \frac{480}{y^4}\right)\right],$$

where $y = aT$. We can combine equations 2-9, 2-9a, and 2-5a to write

$$(2\text{-}15) \qquad \frac{dT}{dQ} = -\frac{T}{Q}\cdot\frac{1}{1+mT}.$$

Then

$$(2\text{-}16) \qquad MC_u = r\frac{dU}{dQ} = r\frac{dU}{dT}\frac{dT}{dQ}.$$

In our computations that include user cost (chapter 3 appendix), we use equation 2-13 for the aggregate user cost and equations 2-14 through 2-16 for the marginal user cost.

Growth in Traffic Loadings

In addition to the growth of all-vehicle traffic, taken into account in our analysis of user cost, most roads are subject to growth in heavy-vehicle traffic. How would this affect our analysis? The answer is that most formulas remain unchanged if we use the average traffic loadings $\overline{Q}$ over the life of the original pavement in place of the constant traffic loadings Q, but that marginal maintenance cost is lowered somewhat. (Of course, uncertainty in the future rate of growth in traffic loadings may also make it difficult to forecast $\overline{Q}$ at the time the investment decision must be made.) We analyze the case where annual vehicle loadings begin at Q_0 and increase at rate f.

To do this, we must separately consider two components of traffic: a baseline growth path Q_0e^{ft} and a small constant increment q per year to that baseline path. It is q whose marginal effect on maintenance cost we wish to determine. Cumulative loadings (in esals) in the right-hand lane at time t are

$$X(t) = \lambda \int_0^t (Q_0 e^{ft'} + q)\, dt' = \lambda \left[Q_0 \frac{e^{ft} - 1}{f} + qt \right]. \tag{2-19}$$

The marginal cost is then the partial derivative of cost with respect to q, evaluated at $q = 0$.

The time T to the first overlay is the time it takes for pavement quality to reach the trigger value. Using the same procedure as in equations 2-3a and 2-5a but with λQt replaced by $X(t)$ implies that

$$X(T)e^{mT} = N. \tag{2-5b}$$

From equation 2-19, $X(T) = \lambda(\overline{Q} + q)T$, where

$$\overline{Q} = \frac{1}{T} \int_0^T Q_0 e^{ft} dt = Q_0 \frac{e^{fT} - 1}{fT} \tag{2-20}$$

is the average traffic level over the first cycle when $q = 0$. Hence

$$T = \frac{N}{\lambda(\overline{Q} + q)} e^{-mT} \tag{2-21}$$

is the equation that replaces equation 2-5a for determining T. When $q = 0$, equation 2-21 is identical to equation 2-5a except that Q is replaced by $\overline{Q}$. Furthermore, substituting equation 2-20 into equation 2-21, and differentiating both sides with respect to q, we can solve for

$$\left(\frac{dT}{dq}\right)_{q=0} = -\frac{\lambda}{N} \cdot \frac{T^2 e^{mT}}{mT + (Q_0 e^{ft}/\overline{Q})}. \tag{2-22}$$

The cost of an overlay is altered by the fact that the overlay must now be thicker in order to bring the pavement to a strength that, at the new starting traffic level Q_0e^{fT}, will give it the same lifetime as before. As a simple approximation, this extra thickness ΔD is given by a power law derived from the AASHO road test:

$$\left(\frac{D + \Delta D + 1}{D + 1}\right)^{A_1} = \frac{Q_0 e^{fT}}{Q_0},$$

which yields

$$\Delta D = (D + 1)[\exp(fT/A_1) - 1], \tag{2-23}$$

where A_1, from table 2-1, is 5.041 for rigid and 7.761 for flexible pavements. If the cost of extra thickness is $k_2W(\Delta D)$ as in equation 2-10, then the new overlay cost is

$$C' = C\{1 + (k_2/k_m)(D + 1)[\exp(fT/A_1) - 1]\}. \tag{2-24}$$

For notational simplicity, we omit the explicit reminder that C and C' depend on W. For typical pavements, C' is higher than C by 6–20 percent.

We are now in a position to recalulate the marginal cost of maintenance:

$$MC_m = r\frac{\partial M}{\partial Q} = -\frac{\alpha C'}{T^2}\left(\frac{dT}{dq}\right)_{q=0}, \tag{2-25}$$

similar to equation 2-7. Inserting equation 2-22 into equation 2-25 yields

$$MC_m = \alpha\,\gamma\,(MC_m^0)\left(\frac{C'}{C}\right), \tag{2-26a}$$

where

$$\gamma = \frac{Be^{mT}}{1 + BmT}, \tag{2-26b}$$

and where

$$B = \frac{\overline{Q}}{Q_0e^{fT}} = \frac{1 - e^{-fT}}{fT}$$

is the ratio of traffic level averaged over the cycle to that at the end of the cycle. Note the similarities between equations 2-26a–b and equations 2-9a–b.

For parameter values typical of heavy-duty rigid freeway pavements ($m = 0, f = 0.04, T = 20, W = 3, D = 10, k_2 = 11{,}718.3, k_m =$ 169,267.2—see table 3-1 for k_2 and k_m values), the two factors causing MC_m to differ from $\alpha(MC_m^0)$, its value in the no-traffic growth case, are $\gamma = 0.688$ and $(C'/C) = 1.13$; their product is 0.78. This indicates that the net effect of traffic growth would be to lower MC_m on such a road by about 22 percent compared with what it would be with constant traffic loadings at the same average value.

CHAPTER THREE

Pavement Wear and Road Durability

BUILDING THE ROAD SYSTEM to optimal durability and replacing current road user taxes with marginal-cost taxes based on pavement wear represents a dramatic change in the nation's highway policy. To estimate the effect of the new policy on the road budget and on suppliers and demanders of freight transport, we use an equilibrium pricing and investment model that links the pricing and investment equations described in the previous chapter with equations describing shippers' demand for truck transport and trucking firms' choice of truck type.

The model, summarized in figure 3-1, operates in a simple iterative scheme incorporating two component models. In the investment component, optimal durabilities are determined by minimizing the sums of capital and maintenance costs for all road classifications.[1] The corresponding marginal costs, calculated on a per-axle-passage basis, are then charged to truckers. In the demand component, truckers respond to these revised user charges by shifting to trucks with more axles, thereby reducing their traffic loadings, and shippers respond by shifting some of their business to or from railroads. On the basis of these predicted loading changes, we make new estimates of traffic loadings for all road classifications. We then recalculate optimal durabilities and marginal costs to be charged to truckers based on their new traffic loadings. The entire process is repeated until the predicted loading changes are negligible. Once we reach this equilibrium, we calculate the resulting effects on the road budget and on truckers and shippers.

Investment Component

As just noted, optimal durability is achieved in our calculations by minimizing the sum of capital and maintenance cost. It is convenient to

1. In the text, we exclude the user costs that result from deterioration in pavement quality because, as already explained, there is so much uncertainty about their magnitude. We show in an appendix how including them might affect our findings.

Figure 3-1. *Equilibrium Pricing and Investment Model*

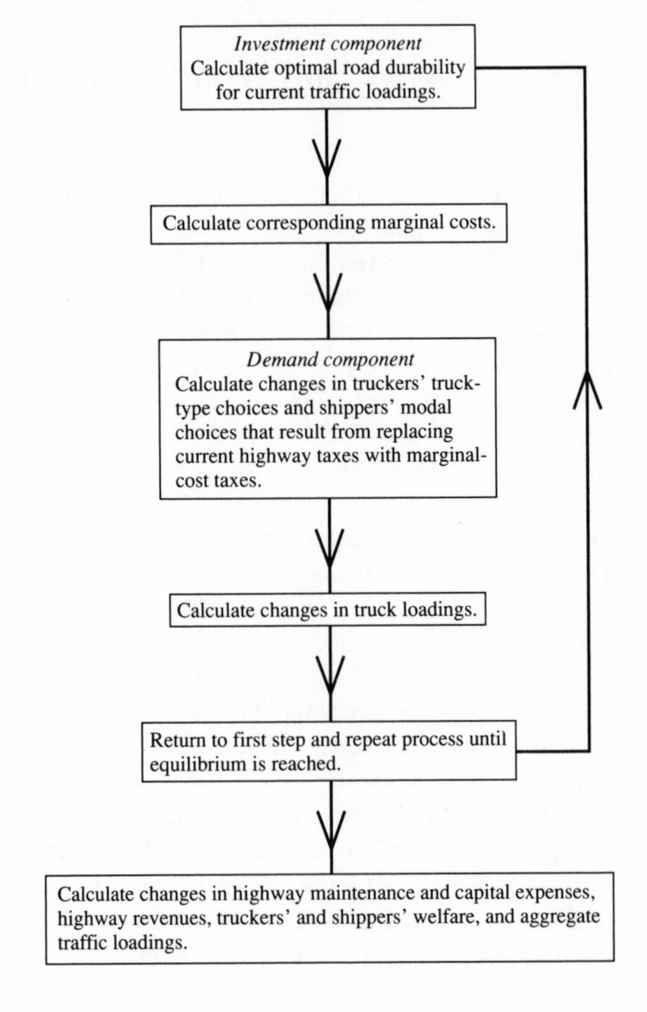

define total pavement costs (*TPC*) as the portion of capital and maintenance cost affected by pavement thickness, expressed per lane. From equations 2-6 and 2-10 in chapter 2 , we can write

$$TPC = k_2D + k_m \cdot \frac{1}{[e^{rT(D)} - 1]},$$

where k_2 is the pavement construction cost per lane-mile per unit of road thickness, D is durability (that is, road thickness), k_m is the resurfacing cost per lane-mile, r is the interest rate, T is the time between resurfacing (which depends on D), and e denotes the exponential function. As explained in chapter 2, durability is measured in inches of pavement thickness for rigid (concrete) pavements, and in an engineering unit called structural number for flexible (asphalt) pavements. The fraction in the above equation is the present value of a unit maintenance expenditure recurring at time intervals $T(D)$.

The relation $T(D)$ is based upon our analysis of the AASHO road test described in the previous chapter. We have empirically measured the other parameters as explained in table 3-1. Resurfacing costs vary by road system and functional classification because traffic disruption and the measures taken to minimize it are far more costly on some types of roads than others.

Optimal durability is determined for each combination of road system, functional class, pavement type, and volume level. There are two road systems (urban and rural), five functional classifications for each (shown in table 3-1),[2] and four surface types (high-type rigid and flexible, and intermediate and low flexible); and within each of the resulting forty combinations, road mileage is further divided into four volume levels. We assume each trucker's loadings to be distributed across road classifications and surface types in the same way as aggregate truck loadings are in his base state. These aggregate distributions are constructed separately for urban and intercity operations in each state, using FHWA estimates of truck vehicle-miles traveled and average esals per truck by functional class. The distribution of truck loadings across surface types within each functional class had to be approximated by the estimated distribution of vehicle-miles traveled by all vehicles.

In order to compare optimal with existing durability, we first estimate the thickness of existing U.S. roads. As noted in chapter 2, because of the wide variation in climate and soil type, we believe we can more accurately describe existing pavements by the time between resurfacing than by their actual thicknesses (for which we lack complete records). We assume times between resurfacing for existing pavements of 13.5 years for rigid freeway and expressway pavements, 20 years for other rigid pavements, and 10 years for flexible pavements. These estimates

2. Two additional functional classifications are urban and rural local roads. We are unable to determine optimal investment for these roads because we lack information on their traffic volumes.

Table 3-1. *Values for Investment Model Parameters*

Parameter	*Description*	*Value*
k_2	Cost of constructing pavement, per lane mile per unit of durability (1982 dollars)	
	Flexible pavement	21,836.1
	Rigid pavement	11,718.3
k_m	Resurfacing costs per lane mile (1982 dollars)	
	Rural	
	Principal arterial—interstate	75,539.2
	Principal arterial—other	54,102.4
	Minor arterial	42,873.6
	Major collector	16,332.8
	Minor collector	18,374.4
	Urban	
	Principal arterial—interstate	169,267.2
	Principal arterial—other freeways	169,267.2
	Principal arterial—other	131,961.6
	Minor arterial	131,961.6
	Collector	131,961.6
r	Real interest rate (percent)	6.0
m	Annual rate of increase in pavement roughness due to aging (percent)	
	Rigid pavement	0.0
	Flexible pavement	4.0
λ	Proportion of loadings occurring in right-hand lane	
	One–three lanes	1.0
	Four–five lanes	0.9
	Six or more lanes	0.7

Sources: Construction costs (k_2) are derived from the average contract price for either Portland cement concrete (rigid pavement) or bituminous concrete (flexible pavement) delivered and spread in place. Cost per unit of delivered material is given in Federal Highway Administration, *Price Trends for Federal-Aid Highway Construction* (Department of Transportation, 1985), p. 2; we have added 15 percent for overhead and deflated to 1982 dollars using the "structures" component of the FHWA highway construction cost index, from Federal Highway Administration, *Highway Statistics, 1986* (Department of Transportation, 1987) p. 58. Resurfacing costs are from the Federal Highway Administration, *Highway Performance Monitoring System Analytical Process,* vol. II, Version 2.0: *Technical Manual* (Department of Transportation, 1986)—our rural parameters are those for "rolling" terrain on p. II-10, and our urban parameters are the average between "built-up" and "outlying" areas on p. II-13, assuming all noninterstates are undivided; we have added disruption costs of 20 percent for urban roads and 10 percent for rural roads, and deflated to 1982 dollars using the "surfacing" component of the FHWA index cited above. The interest rate should represent the alternative real cost of public funds; our choice of 6 percent lies between the historical real rate of return on bonds and real pretax rate of return in the private sector. The increase in the rate of roughness due to aging is discussed in the chapter 2 appendix. The proportion of loadings occurring in the right-hand lane is from the FHWA index cited above, p. II-16.

are based on interviews with highway engineers, a review of design guides, and our own studies of highway test data.[3] We then work backward, using our method for predicting years of pavement life from road thickness and traffic loadings, to estimate the road thickness itself

3. See Kenneth A. Small and Clifford Winston, "Optimal Highway Durability," *American Economic Review,* vol. 78 (June 1988), p. 567, note 15.

Table 3-2. *Bounds on Estimated Road Thicknesses for Existing Roads*

Road category	Durability	
	Lower bound	*Upper bound*
Low[a]	1.44	2.60
Intermediate[a]	1.44	3.35
High-type flexible[a]		
Nonfreeway	2.32	5.36
Freeway	4.64	6.80
High-type rigid[b]		
Nonfreeway	6.00	10.00
Freeway	7.50	12.50

Source: FHWA.
a. Durability is measured by a composite number called a structural number.
b. Durability is pavement thickness measured in inches.

for a given functional class, pavement type, and traffic-level interval. We place upper and lower bounds on these estimates to ensure consistency with the definition of that pavement type.[4] The bounds on road thickness, given in table 3-2, are based on definitions of the roads used by the FHWA, supplemented by additional information provided by the FHWA. In the optimal durability calculations, rigid pavement is constrained to be at least 2.0 inches thick and flexible pavement to have a structural number of at least 1.44, representing practical minimum thicknesses. (Experimentation showed that, in fact, the bounds on initial durability had only a small effect on the results, and the bounds on optimal durability had virtually no effect on the results.)

Table 3-3 summarizes the results for current traffic loadings (that is, when only the first step in figure 3-1 is carried out). For example, we estimate the maintenance cost on an existing rural interstate to be 1.48 cents per standard axle passage per mile, a figure very close to the average maintenance cost calculated by the National Cooperative Highway Research Program.[5]

4. These definitions are summarized in FHWA, *Highway Statistics, 1986* (Department of Transportation, 1987), p. 115, table HM-12, note 3. We obtained more complete descriptions from FHWA engineers.

5. Transportation Research Board, National Cooperative Highway Research Program, "Relationships between Vehicle Configurations and Highway Design" (Washington, D.C.: National Research Council, November 5, 1986); their estimate is 1.6 cents per esal-mile.

Table 3-3. *Average Durability and Marginal Maintenance Cost by Road Classification under Current Traffic Levels*[a]

	Durability				Marginal maintenance cost (cents per esal-mile)	
	Rigid pavement[b]		Flexible pavement[c]			
Road classification	*Current*	*Optimal*	*Current*	*Optimal*	*Current invest-ment*	*Optimal invest-ment*
Rural						
Principal arterial						
Interstate	9.52	11.35	5.26	6.43	1.48	0.46
Other	7.79	8.67	3.90	4.91	4.38	1.13
Minor arterial	6.52	6.59	3.30	4.13	10.02	2.60
Major collector	6.00[d]	3.74[d]	2.46	2.69	16.49	9.96
Minor collector	6.00[d]	3.37[d]	2.18	2.42	31.18	16.09
Local	6.00[d]	6.00[e]	2.18	2.18	101.30	101.30
Urban						
Principal arterial						
Interstate	10.07	13.52	5.56	7.69	2.38	0.33
Other freeways	9.21	11.81	4.97	6.79	4.32	0.61
Other	7.92	10.04	4.21	6.04	10.92	0.87
Minor arterial	6.78	7.50	3.22	4.79	33.92	3.23
Collector	6.00[d]	4.97[d]	2.51	3.73	125.45	13.66
Local	6.00[d]	6.00[e]	2.51	2.51[e]	40.92	40.92

Source: Authors' calculations based on loading levels and distributions derived from disaggregate average daily traffic volume for the above road classifications provided by the FHWA.

a. The marginal costs presented are averaged across low, intermediate, and high-type pavements, and across four volume ranges.

b. Durability is pavement thickness measured in inches.

c. Durability is measured by a composite index, called a structural number, reflecting pavement, base, and sub-base thicknesses in inches with weights 0.44, 0.14, and 0.11, respectively.

d. There are relatively few road-miles of collector or local roads with rigid pavement. Durability under current investment is assumed for such roads to be equal to the minimum of the range defining "high-type rigid" pavements, although this leads to extremely long lifetimes.

e. We are unable to determine optimal investment for local roads because we lack information on their traffic volume. Thus, their durability is assumed the same under current and optimal investment.

As can be seen in the table, optimal investment leads to dramatic changes in road durability and marginal maintenance cost. For most road classifications, including all the arterials, optimal investment results in more durability, hence lower marginal maintenance costs, than does current practice.[6] Interestingly, there is a historical parallel for road design based on overly optimistic engineering forecasts. A brief boom in "plank roads" during the 1840s, between the era of turnpikes and that

6. Two of us have argued elsewhere that the primary reasons for this are past reliance upon statistically flawed design equations and failure to incorporate economic optimization. See Small and Winston, "Optimal Highway Durability," p. 568.

of canals, was based on widely disseminated estimates that the wooden planks used in them would last eight to twelve years. Actual lives were about half that, with disastrous financial consequences for the hundreds of private road companies who built them.[7]

The changes in marginal maintenance costs have important implications for user charges. Tables 3-4 and 3-5 compare, for urban and intercity trucks, respectively, three systems of user charges: current taxes, a marginal maintenance-cost pricing system at current investment, and a marginal maintenance-cost pricing system at optimal investment.[8] (Hereafter in this chapter we sometimes use the abbreviated description "marginal cost" for MC_m.) Both tables compute the resulting charges per vehicle-mile for the trucks depicted in figure 3-2.[9]

Current pricing, as detailed in chapter 1, consists largely of fuel taxes, supplemented by registration fees, special excise taxes, and (in a few states) taxes graduated by truck weight and distance traveled. The tables demonstrate that under current pricing, user charges rise only slowly with gross vehicle weight (for a given truck type), and that, for a given gross vehicle weight, the charges rise with the number of axles. Marginal-cost taxes, in contrast, would rise steeply with gross vehicle weight, for a given truck type, and *fall* as the number of axles rises.

The economic perversity of current practice is illustrated by some specific cases. An urban two-axle van (SU2) with gross vehicle weight of 33,000 pounds pays on average 3.01 cents per vehicle-mile under current policy rather than its very high marginal maintenance cost of 23.77 cents per vehicle-mile. On the other hand, a lightly loaded urban

7. Daniel Klein and John Majewski, "Private Profit, Public Good, and Engineering Failure: The Plank Roads of New York," Working Paper 8813 (Institute for Humane Studies, George Mason University, October 1988).

8. A truck was considered to be in urban operation if it served a local jurisdiction or met the following conditions: 100 percent of the trips were within a fifty-mile radius of the base of operations and an interstate jurisdiction was not served.

9. To calculate the weight on each axle for a given load, we used Federal Highway Administration estimates of the distribution of gross vehicle weight over axles. These estimates vary by truck type and by how fully loaded the vehicle is. For example, the percentage of gross vehicle weight on each axle of a single-unit two-axle truck (SU2) is: 38 percent on the first axle and 62 percent on the second if it is fully loaded, 39.3 percent on the first axle and 60.7 percent on the second if it is partially loaded, and 43.1 percent on the first axle and 56.9 percent on the second if it is empty. We applied a fourth-power approximation to determine loadings for axle weights different from 18,000 pounds.

Figure 3-2. *Truck Types*

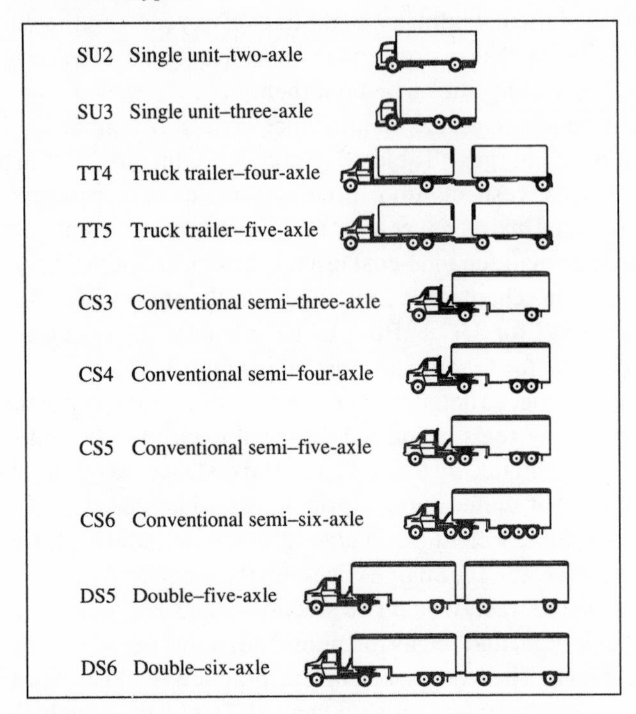

tractor-trailer (CS5) with the same gross vehicle weight faces a 4.07 cent charge under current user taxes, yet causes relatively little damage as evidenced by its marginal maintenance cost of only 1.20 cents.

The bottom panels of tables 3-4 and 3-5 show that the road-wear charges fall dramatically when they are accompanied by optimal investments in road durability. The mainstay of interstate trucking, a tractor-trailer (CS5) with gross weight 80,000 pounds, would on average see its marginal-cost charges lowered from 14.46 cents to 3.09 cents a mile (table 3-5). (Its current taxes are 4.96 cents a mile.) Thus current policy not only creates inefficient incentives for truckers making choices of load and equipment, but also deprives many truckers and shippers of the benefits from lower road taxes that could accompany optimal investment.

Demand Component

Replacing current user taxes with marginal-cost taxes would lead to several responses by truckers, some of which are more important and

Table 3-4. *Current Truck Taxes and Marginal-Cost Taxes for Urban Trucks with Current Traffic Levels*
1982 cents per vehicle-mile

Vehicle type	*Gross vehicle weight (thousands of pounds)* 26	33	55	80	105
			Current taxes[a]		
SU2	2.52	3.01	4.22	...	...
SU3	3.88	4.38	5.61	7.43	...
TT4	...	...	5.08	6.93	8.01
TT5	...	...	5.94	7.79	8.87
CS3	3.21	3.69	4.95	6.81	...
CS4	...	4.04	5.31	7.17	8.25
CS5	...	4.07	5.34	7.19	8.28
CS6	...	4.09	5.36	7.21	8.29
DS5	...	...	6.01	7.85	8.96
DS6	...	...	6.06	7.90	9.01
			Marginal maintenance-cost taxes: current investment		
SU2	9.16	23.77	183.38	...	...
SU3	2.07	5.37	41.43	125.43	...
TT4	...	...	23.67	105.94	314.39
TT5	...	...	9.18	41.07	121.87
CS3	2.30	6.16	47.54	212.78	631.43
CS4	...	2.93	22.61	101.19	300.30
CS5	...	1.20	9.22	41.26	122.44
CS6	...	0.71	5.45	24.42	72.45
DS5	...	1.30	10.04	44.92	133.31
DS6	...	0.81	6.22	27.83	82.58
			Marginal maintenance-cost taxes: optimal investment		
SU2	3.59	9.32	71.89	...	...
SU3	0.81	2.11	16.24	72.69	...
TT4	...	...	9.28	41.53	123.25
TT5	...	...	3.60	16.05	47.78
CS3	0.90	2.42	18.63	83.41	...
CS4	...	1.15	8.86	39.67	117.72
CS5	...	0.47	3.61	16.17	48.00
CS6	...	0.28	2.14	9.57	28.40
DS5	...	0.51	3.93	17.62	52.26
DS6	...	0.32	2.44	10.91	32.37

Sources: Current taxes from American Trucking Association, based on annual miles of travel for trucks in our data base. Marginal maintenance-cost taxes from authors' calculations.

a. Figures include fuel taxes, registration fees that vary by weight, special excise taxes, and weight-distance or ton-mile taxes. The CS6 and DS6 vehicle types include combinations with six or more axles.

Table 3-5. *Current Truck Taxes and Marginal-Cost Taxes for Intercity Trucks with Current Traffic Levels*
1982 cents per vehicle-mile

Vehicle type	Gross vehicle weight (thousands of pounds)				
	26	*33*	*55*	*80*	*105*
			Current taxes[a]		
SU2	1.95	2.24	2.91	...	...
SU3	3.25	3.55	4.23	5.31	...
TT4	...	...	3.57	4.67	5.27
TT5	...	...	4.41	5.51	6.11
CS3	2.51	2.77	3.49	4.59	5.19
CS4	...	3.14	3.84	4.94	5.54
CS5	...	3.16	3.86	4.96	5.56
CS6	...	3.17	3.86	4.97	5.67
DS5	...	...	4.44	5.54	6.15
DS6	...	...	4.46	5.56	6.17
			Marginal maintenance-cost taxes: current investment		
SU2	3.21	8.33	64.26	...	...
SU3	0.73	1.88	14.52	64.98	...
TT4	...	...	8.29	37.13	110.18
TT5	...	...	3.22	14.39	42.71
CS3	0.81	2.16	16.66	74.57	221.28
CS4	...	1.03	7.92	35.46	105.24
CS5	...	0.42	3.23	14.46	42.91
CS6	...	0.25	1.91	8.56	25.39
DS5	...	0.46	3.52	15.74	46.72
DS6	...	0.28	2.18	9.75	28.94
			Marginal maintenance-cost taxes: optimal investment		
SU2	0.69	1.78	13.74	...	...
SU3	0.16	0.40	3.10	13.89	...
TT4	...	...	1.77	7.94	23.55
TT5	...	...	0.69	3.08	9.13
CS3	0.17	0.46	3.56	15.94	47.29
CS4	...	0.22	1.69	7.58	22.49
CS5	...	0.09	0.69	3.09	9.17
CS6	...	0.05	0.41	1.83	5.43
DS5	...	...	0.75	3.37	9.99
DS6	...	...	0.47	2.08	6.19

Sources: See table 3-4.
a. Figures include fuel taxes, registration fees that vary by weight, special excise taxes, and weight-distance or ton-mile taxes. The CS6 and DS6 vehicle types include combinations with six or more axles.

more likely than others. Most important, truckers would tend to switch their heavy loads to truck types with more axles in order to lower their user charges. Truckers would also adjust load sizes, dispatch schedules, and fleet sizes to balance road taxes and economies from larger loads and fully utilized capacity. Less significant responses would be routing shifts to balance road taxes and travel time, and relocation of terminals to balance road taxes and other network considerations. Marginal-cost taxes would also affect shippers' and railroads' profits as truck rates change and shippers alter their modal choices.

In this section we analyze the most important response by truckers: type shifting, or changing the type of truck in which a given cargo is carried. We also estimate the further changes in railroad and shipper profits caused by modal shifts. The other responses by truckers will be considered later.

To analyze type shifting, we develop and estimate a disaggregate econometric model of a trucker's choice of truck type. The trucker is assumed to select, from among the ten truck types displayed in figure 3-2, the one that maximizes profits for the maximum load to be transported. The choice is influenced by the characteristics of the truck types such as capital cost, operating cost, volume, durability, and maneuverability; regulatory constraints such as weight limits and illegality of a particular truck type in a particular state; and characteristics of the trucker's operation such as maximum load size and kinds of commodities transported.

Each of these influences is represented by one or two variables, as listed in table 3-6. We begin with the truck's characteristics. Large capital and operating costs should decrease the likelihood of a truck type's being selected, whereas a large volume of cargo space should have a positive influence. We represent a truck's susceptibility to wear by its annual miles divided by its number of axles; an increase in this variable should have a negative influence. Lack of maneuverability is represented by two "dummy" variables, one indicating whether a truck is a combination vehicle, the other indicating whether it has a trailer; all else equal, these features should discourage choosing that type.

Of the regulatory constraints and operating characteristics, the maximum allowable weight per axle for a given truck type should have a positive influence on its choice because of the additional flexibility it provides; whereas if a given truck is illegal under the particular load carried or if the truck type itself is illegal in the trucker's state, the probability of choosing that type should be substantially reduced.

Table 3-6. *Multinomial Logit Truck Type Choice Estimates*

Explanatory variable	*Coefficient*[a] *Urban operations*	*Coefficient*[a] *Intercity operations*
Truck characteristics		
Capital cost (thousands of dollars)[b]	−0.119	−0.065
	(0.045)	(0.012)
Operating cost (thousands of dollars a year)[c]	−0.764	−0.343
	(0.266)	(0.068)
Annual vehicle miles ÷ number of axles	−0.370	−0.168
(thousands of miles per axle)	(0.136)	(0.038)
Volume of cargo space (cubic feet)	0.0022	0.0008
	(0.0016)	(0.0003)
Combination dummy (1 if CS3, CS4, CS5,	−8.256	−4.412
CS6 alternative; 0 otherwise)	(2.466)	(0.501)
Trailer dummy (1 if TT4, TT5, DS5, DS6;	−10.06	−7.593
0 otherwise)	(3.379)	(0.759)
Regulatory constraints		
State weight limit per axle	0.808	0.766
(thousands of pounds per axle)[d]	(0.246)	(0.078)
Illegal truck dummy (1 if load plus truck's	−3.502	−1.839
empty weight exceed weight limit or if	(1.245)	(0.396)
type is illegal in the trucker's base state;		
0 otherwise)[e]		
Characteristics of trucker's operation		
Maximum load (thousands of pounds,	0.119	0.152
defined for five-axle and six-axle	(0.044)	(0.012)
alternatives)		
Commodity dummy (1 if garbage or	2.810	...
cement truck; 0 otherwise, defined for	(3.519)	
single-unit alternatives)		

Source: Authors' calculations.

a. Standard errors in parentheses; for the urban model log likelihood at zero is −322.4; log likelihood at convergence is −34.77; 140 observations. For the intercity model log likelihood at zero is −1105.0; log likelihood at convergence is −246.7; 480 observations.

b. Capital cost is vehicle's prevailing purchase price. Prices of chassis are from *The Blue Book for Heavy Trucks* (Chicago: MacLean Hunter Market Reports, 1982); prices for bodies and trailers are from various truck dealers.

c. Operating cost includes fuel costs and all taxes. Truck taxes are from the American Trucking Association; price of fuel is from the American Petroleum Institute; and fuel economy is from the Truck Inventory and Use Survey.

d. State weight limits are from the American Trucking Association.

e. Ideally, the variable definition should be expanded to account for the potential illegality of a vehicle for all states in which the vehicle was driven. Unfortunately, the sample did not reveal all states in which the vehicle was operated.

Finally, we expect larger loads to increase the likelihood of selecting the largest trucks; and in urban operations we expect certain commodities that are carried on local streets, namely cement and garbage, to increase the likelihood of selecting the smaller, more maneuverable trucks.

We assume that the load carried is exogenous, that is, not under the carrier's control. The assumption does not seem to have much effect on the results.[10] Operating costs and a truck's legality are determined by a truck's gross vehicle weight (empty weight plus maximum load) as well as its axle configuration.[11]

The probabilistic choice model is estimated from a sample contained in the 1982 Truck Inventory and Use Survey, part of the U.S. Census of Transportation. This survey includes truck type, annual vehicle-miles, fuel economy, maximum and average loads, and commodities transported. Separate models are estimated for urban and intercity operations.[12] Corrections are made in the estimation procedure for the nonrandom sampling process.[13]

Probabilistic choices are represented by the multinomial logit model;[14] alternative models were considered but were not found to represent

10. A valid statistical test of the assumption requires specifying and estimating a joint choice model of truck type and load. Such a model could not be estimated here for reasons discussed at the end of this chapter. A suggestive statistical test was carried out by using a reduced-form equation to predict load size and using the predicted load instead of the actual load in forming the independent variable. Although the coefficient on "maximum load" was slightly higher when the predicted rather than the actual load was used, there was no statistically significant difference, and the coefficients of the other variables were virtually unaffected.

11. Because we use maximum load (that is, the maximum load that the truck carries), we assume, based on Bureau of Economics, *Empty/Loaded Truck Miles on Interstate Highways During 1976* (Interstate Commerce Commission, 1977), that trucks travel 80 percent of their mileage with a full load and 20 percent of it empty. (Truckers' reports of average loads in our data appear to be less reliable than their reports of maximum load.) We attempted to obtain empty/loaded distributions by vehicle type. Based on a preliminary sample, the Federal Highway Administration estimated that single-unit trucks (SU2, SU3) traveled 60 percent of their mileage with a full load and 40 percent of it empty, while the other vehicle types traveled 75 percent of their mileage full and 25 percent of it empty. Using these assumptions instead of those based on the ICC study changes the estimated net benefits from our pricing and investment policy by only 4 percent.

12. The definition of urban and intercity operations is given in note 8. Of course, most trucking *firms* engage in both types of operations.

13. The sample was stratified by truck type and geographic location (state of truck's origin). To correct for both sources of nonrandomness, our model was estimated using the weighted exogenous sample maximum likelihood (WESML) method, with sampling weights for each observation equal to the percent of the truck population from state S selecting truck type i, divided by the percent of the sample truck population from state S selecting truck type i. For a discussion of the appropriateness of this estimator under choice-based sampling with geographical stratification, see Daniel McFadden, Clifford Winston, and Axel Boersch-Supan, "Joint Estimation of Freight Transportation Decisions under Nonrandom Sampling," in Andrew F. Daughety, ed., *Analytical Studies in Transport Economics* (Cambridge University Press, 1985). These sampling weights were also used in all calculations using the sample and in all statistical tests of the model. To simplify exposition, explicit mention of these weights is not repeated in our discussions.

14. The multinomial logit choice probabilities are given by

improvements.[15] Our parameter estimates are presented in table 3-6. The coefficients all have the expected signs and are generally statistically reliable. The negative coefficients for the combination and trailer dummies indicate that truckers, especially those with urban operations, have a strong preference for single-unit trucks if all else is equal.

Because our policies employ user charges to induce shifts to vehicles with more axles and higher capital costs, the capital and operating cost parameters are especially important to our model. A quantitative sense of their plausibility can be obtained by using them to calculate the real implicit discount rates that operators are using in their capital decisions if our model correctly describes their behavior. We find these rates to be 6.35–10.32 percent for urban truckers and 5.14–9.58 percent for intercity truckers.[16] These rates are plausible because they closely bracket historical pretax industrial real rates of return;[17] and because they are lower than recent estimates of consumers' implicit discount rates (as one would expect since producers are presumably less myopic and have greater access to capital markets than consumers).[18]

$$\text{Prob}_i = \exp(\beta X_i) / \sum_{j=1}^{10} \exp(\beta X_j),$$

where Prob_i is the probability of selecting truck type *i*, parameters are denoted by β, and explanatory variables denoted by X. A questionable characteristic of this model is that it assumes that unobserved effects are uncorrelated across alternatives. We tested the validity of this assumption applying a test suggested by Kenneth A. Small and Cheng Hsiao, "Multinomial Logit Specification Tests," *International Economic Review*, vol. 26 (October 1985), pp. 619–27, using subsamples of the ten alternative truck types selected at random and composed of specified classes of vehicles (for example, all CS truck types). In all cases, the null hypothesis of a multinomial logit structure was not rejected at reasonable confidence levels.

15. We ordered the truck type alternatives by number of axles and estimated the ordered logit model developed by Kenneth A. Small, "A Discrete Choice Model for Ordered Alternatives," *Econometrica*, vol. 55 (March 1987), pp. 409–24. This model fit the data poorly. We also considered various nested logit structures, but encountered estimation difficulties because of the small number of chosen DS types in our sample.

16. The discount rates are values for which a one dollar increase in capital costs and a one dollar increase in the present value of operating costs have the same effect on type choice. The calculation assumes a specified duration of vehicle ownership. Typical ownership durations are eight to ten years for urban firms and six to seven years for intercity firms; the ranges reported in the text correspond to these lengths of ownership.

17. See Federal Trade Commission, *Quarterly Financial Report for Manufacturing, Mining, and Trade Corporations* (FTC, various issues).

18. Recent estimates of consumers' implicit discount rates in the purchase of durables

This choice model is used to calculate the change in truckers' welfare, ΔW, from instituting a policy such as marginal-cost user charges. Although we refer to the welfare gain as being that of truckers, much of it will actually be passed on to shippers and consumers through adjusted prices. This welfare change is calculated for each member of our sample from the formula

$$\Delta W = -\frac{1}{\beta_1}\left[\ln \sum_{i=1}^{10} \exp(\beta X_i)\right]_{\beta X_i^0}^{\beta X_i^1},$$

where ln denotes the natural logarithm, β denotes the set of coefficient estimates from the demand model, X denotes the set of explanatory variables, i is a truck type, the superscripts for X denote the value of the variables before and after the policy change, and β_1 is a conversion factor to put the results in monetary units.[19] We calculate the average welfare change for the sample and multiply it by the total number of trucks in the United States in order to estimate changes in aggregate truckers' welfare.

Because the change in taxes will influence trucking rates, shippers and railroads will also be affected as shippers move some shipments from rail to truck and others from truck to rail. These modal shifts are estimated using a previously published disaggregate modal split model for freight.[20] They generate two additional welfare changes, both of which are small. The first is the change in "modal surplus," calculated by the usual Harberger "rule of a half," applied to the change in trucking ton-miles for each commodity.[21] The second is the change in rail profits,

range from 15 percent to 30 percent; see Jerry A. Hausman, "Individual Discount Rates and the Purchase and Utilization of Energy-Using Durables," *Bell Journal of Economics*, vol. 10 (Spring 1979), pp. 33–54; and Fred Mannering and Clifford Winston, "A Dynamic Empirical Analysis of Household Vehicle Ownership and Utilization," *Rand Journal of Economics*, vol. 16 (Summer 1985), pp. 215–36.

19. Kenneth A. Small and Harvey S. Rosen, "Applied Welfare Economics with Discrete Choice Models," *Econometrica*, vol. 49 (January 1981), pp. 105–30. Because ΔW is a welfare measure, the term inside the brackets is measured in abstract units of "utility." The parameter β_1, obtained from Roy's Identity as equal to the operating cost coefficient, converts the expression in brackets into a welfare change in thousands of dollars a year per truck.

20. Clifford Winston, "A Disaggregate Model of the Demand for Intercity Freight Transportation," *Econometrica*, vol. 49 (July 1981), pp. 981–1006.

21. The expression is 1/2 (change in ton-miles) x (change in price per ton-mile). For a derivation of this formula in this context see Kenneth A. Small and Clifford Winston, "Welfare Effects of Marginal-Cost Taxation of Motor Freight Transportation: A Study

calculated as the increase in rail shipments multiplied by the difference between rail's price and its marginal cost.[22]

Finally, the change in tax revenues is calculated as the difference between the (1982) tax revenues and what revenues would be under marginal-cost taxation.[23] We assume that the marginal costs calculated from the investment model (for each combination of road system, functional class, pavement type, and traffic volume) replace the current (1982) fuel taxes, registration taxes, and weight-distance taxes. Those few fees that are independent of gross vehicle weight or vehicle-miles traveled are assumed to remain in place and to approximately cover administrative expenses.

Findings

The economic effects of applying marginal-cost taxes and building highways to optimal durability are summarized in table 3-7. These findings apply to the steady state, after the policies have been in effect for many years. There will be a lengthy transition during which capital expenses are incurred to rebuild the highway network but the maintenance savings are not yet enjoyed; we have not estimated the interim effects during this period.

---

of Infrastructure Pricing," in Harvey S. Rosen, ed., *Studies in State and Local Finance* (University of Chicago Press, 1986).

22. Estimates from the Association of American Railroads, corroborated by existing rail econometric cost models, conservatively suggest that the difference between rail's price and its marginal cost is 0.5 cents per ton-mile. The change in net ton-miles for each member of our sample is equal to the elasticity of demand for trucking services multiplied by the percentage change in trucking prices caused by the tax, multiplied by truck's original ton-miles. This entire change is assumed due to shifts to or from rail. Calculations are carried out by commodity group; for each commodity the elasticity of demand for trucking services is obtained from Winston's freight demand model, "A Disaggregate Model of the Demand for Intercity Freight Transportation." The average price elasticity based on this demand model is 0.7. Calculations were also performed assuming a uniform price elasticity for all commodities of 1.0 with little effect on the results.

23. For every truck type choice in the sample, the tax burden that would be incurred if the truck were chosen is multiplied by the probability of that truck's being chosen. An average expected tax is then calculated for the sample and multiplied by the number of trucks in the population to obtain aggregate revenues for 1982 taxes and for marginal-cost taxes.

Table 3-7. *Economic Effects of Optimal Pricing and Investment*
Change, relative to current practice, in billions of 1982 dollars except as noted

Item	*Effect*
Investment costs	
Maintenance savings[a]	9.428
Capital savings	−1.276
Total savings	8.152
Trucking firms' and shippers' welfare	
Intercity operations	0.720
Urban operations	−0.586
Total	0.134
Revenues[b]	−0.574
Modal Shifting	
Modal surplus	0.029
Rail profits	0.011
Total welfare	7.752
Change in loadings (percent)	−38.120

Source: Authors' calculations.
a. Includes $1.25 billion in disruption cost savings to motorists.
b. Revenues include gains or losses from modal shifting.

The policy leads to substantial gains in net welfare in the steady state, totaling $7.75 billion a year (in 1982 dollars). The bulk of the benefits consists of annual savings of more than $9 billion in maintenance costs (including $1.25 billion in disruption costs) or more than 75 percent of current maintenance costs, which now approach $12 billion. Surprisingly, these savings are achieved with only a $1.3 billion increase in annualized capital costs. Annual highway revenues are reduced slightly, about $600 million. The net result is a $6.3 billion improvement each year in the public sector's long-run cash flow, a benefit that is all the more significant because of the difficulty of reducing the U.S. budget deficit and because of the substantial inefficiencies believed to be associated with most sources of tax revenue. The source of the welfare gain is the 38 percent decrease in truck loadings (esal-miles) resulting from our policy, almost all of which is due to shifts among truck types.

Surprisingly, this policy results in little redistribution among broadly defined interest groups—truckers, shippers, railroads, and highway contractors. In fact, each of these groups is likely to be better off, even assuming no return of the public-sector benefits to them. Truckers' user charges go up for some roads and commodities and down for others, but

Table 3-8. *Current and Optimal Capital and Maintenance Costs and Durability, by Road Classification*[a]

Billions of dollars except as noted

	Capital costs			Maintenance costs			Durability			
							Rigid pavement[b]		Flexible pavement[c]	
Road classification	*Current*	*Optimal*	*Change*	*Current*	*Optimal*	*Change*	*Current*	*Optimal*	*Current*	*Optimal*
Rural										
Principal arterial										
Interstate	0.894	1.013	0.119	0.551	0.134	−0.417	9.52	10.41	5.26	6.09
Other	1.091	1.267	0.176	0.854	0.196	−0.658	7.79	7.93	3.90	4.65
Minor arterial	1.330	1.542	0.212	1.074	0.256	−0.818	6.52	5.76	3.30	3.88
Major collector	2.419	2.453	0.034	0.808	0.463	−0.345	6.00[d]	3.47[d]	2.46	2.53
Minor collector	1.073	1.024	−0.049	0.441	0.204	−0.237	6.00[d]	2.25[d]	2.18	2.10
Urban										
Principal arterial										
Interstate	0.410	0.512	0.102	0.627	0.056	−0.571	10.07	12.20	5.56	7.14
Other freeways	0.230	0.283	0.053	0.335	0.032	−0.303	9.21	10.66	4.97	6.38
Other	0.873	1.143	0.270	2.483	0.162	−2.321	7.92	8.81	4.21	5.67
Minor arterial	0.793	1.001	0.208	1.989	0.154	−1.835	6.78	5.96	3.22	4.20
Collector	0.524	0.676	0.152	1.267	0.120	−1.147	6.00[d]	3.92[d]	2.51	3.33
Total	9.637	10.914	1.277	10.429	1.777	−8.652	...	...	...	...

Source: Authors' calculations based on loading levels and distributions derived from disaggregate average daily traffic volume for the above road classifications provided by the FHWA.

a. Local roads are not included.

b. Durability is pavement thickness measured in inches.

c. Durability is measured by a composite index, called a structural number, reflecting pavement, base, and sub-base thicknesses in inches with weights 0.44, 0.14, and 0.11, respectively.

d. With the exeption of arterials, there are relatively few road miles of rigid pavement. Durability under current investment is assumed for rigid collector and local roads to be equal to the minimum of the range defining "high-type rigid" pavements, although this leads to extremely long lifetimes.

the table shows that truckers gain on balance (and since most firms do not restrict themselves to just urban or intercity operations, it would be misleading to view one group of firms as gaining at the expense of another). Shippers gain from the passing through of some portion of these lower user charges. Railroads gain because charges tend to rise on intercity traffic shipped long distances in large quantities; hence their business grows despite the small overall decrease in average user charges for intercity truck operations.[24] (Rail is assumed not to compete on urban operations.) Finally, highway contractors gain business in the short run as roads are being rebuilt to higher durability; only after many years does the less frequent resurfacing reduce the demand for highway repair.

Subsequent tables present more detailed results. Table 3-8 gives a breakdown of current and optimal capital and maintenance costs and of durability by road classification. Increased road durability leads to reduced maintenance costs for all classifications, especially for urban arterials and collectors. Within each classification, these reductions are achieved with only modest increases in long-run annualized capital expenses. Table 3-9 reveals the effect of the policy on type choice, loadings, and vehicle-miles traveled. The most dramatic shift is from the relatively damaging SU2 to truck types with more axles; the overall result, as indicated earlier, is a 38 percent decrease in vehicle loadings.

Type shifts and greater durability explain why truckers as a group are not adversely affected by the policy. As shown in table 3-10, the optimal user charges vary from 0.5 cents to more than $1.00 per axle passage, as compared with current taxes that we calculate to average 1.34 cents per axle passage and do not vary by road classification. Under our policy, road taxes would be lower than they are now on the most heavily traveled interstates and principal arterials, but higher on the less traveled collec-

24. For intercity truck operations, the average cost of a ton-mile decreases, on average. But the trucks whose average costs increase have average loads of 41,095 pounds and average annual vehicle-miles traveled of 71,519. Trucks whose average costs decrease have average loads of only 16,060 pounds and average annual vehicle-miles traveled of only 37,235. Thus, rail experiences a net gain in ton-miles, even though trucking's average cost falls, because truck costs rise on movements that, on average, have large ton-miles and fall on movements that, on average, have fewer ton-miles.

If one considers truck types, it might be expected that trucks would gain ton-miles from rail because the "rail competitive" trucks, for example the five- and six-axle double trailers and combinations, experience cost reductions. But all truck types experience cost increases and decreases, and the final modal distribution of traffic will again depend on the initial ton-miles and price for particular movements.

Table 3-9. *Predicted Distribution of Vehicles, Loadings, and Vehicle Miles Traveled under Current Pricing and Investment and under Optimal Pricing and Investment*
Percent

	Intercity		Urban	
Type	*Current*	*Optimal*	*Current*	*Optimal*
		Vehicles		
SU2	69.36	65.72	91.22	84.58
SU3	6.31	7.07	6.34	9.43
TT4	0.26	0.24	0.33	0.67
TT5	0.13	0.16	0.01	0.02
CS3	2.21	2.27	0.75	1.67
CS4	0.86	0.95	0.24	0.62
CS5	17.15	18.24	0.78	1.13
CS6	3.19	4.77	0.19	1.36
DS5	0.51	0.56	0.13	0.45
DS6	0.02	0.02	0.01	0.07
		Loadings		
SU2	48.16	19.21	91.40	63.29
SU3	0.77	1.25	2.34	9.52
TT4	0.43	0.21	0.44	1.16
TT5	0.18	0.33	0.03	0.10
CS3	2.20	1.72	0.61	2.89
CS4	0.78	0.75	0.23	0.79
CS5	40.22	59.43	3.74	8.23
CS6	5.91	14.94	0.85	11.75
DS5	1.33	2.11	0.30	1.71
DS6	0.04	0.06	0.06	0.56
		Vehicle-miles traveled		
SU2	44.82	40.82	89.32	79.22
SU3	3.14	3.71	5.27	8.36
TT4	0.38	0.26	1.09	1.78
TT5	0.19	0.22	0.02	0.03
CS3	2.05	1.87	0.80	1.74
CS4	0.95	0.93	0.30	0.71
CS5	39.01	38.15	2.23	2.51
CS6	8.46	12.99	0.71	4.64
DS5	0.97	1.03	0.23	0.75
DS6	0.03	0.03	0.04	0.26

Source: Authors' calculations.

Table 3-10. *Marginal-Cost Pavement-Wear Charges, by Road Classification under Optimal Pricing and Investment*[a]
Cents per esal-mile

Road classification	*Marginal-cost user charges*
Rural	
Principal arterial	
Interstate	0.66
Other	1.63
Minor arterial	3.90
Major collector	14.50
Minor collector	36.63
Local[b]	101.30
Urban	
Principal arterial	
Interstate	0.52
Other freeways	0.94
Other	1.35
Minor arterial	7.64
Collector	27.77
Local[b]	40.92

Source: Authors' calculations.
a. Pavement-wear charges are averaged across pavement types.
b. Reflects current durability.

tors and minor arterials. These changes amount to a slight increase in average taxes and costs for urban truckers and a slight decrease for intercity truckers. As shown in table 3-11, the resulting change in net welfare is only a few hundred dollars per truck per year for each type of operation; as a percentage of total trucking costs, it averages to a 0.1 percent net gain.

Table 3-12 presents the effect of our policy on the maintenance costs and tax revenue attributable to each truck type as a share of all such attributable maintenance costs. Under current policy, most vehicles' taxes fall far short of maintenance costs attributable to them, especially the SU2 and CS5. Summing the second column of the table reveals that taxes on intercity trucks cover only 29 percent of their attributable pavement maintenance costs and that urban truck taxes cover only 14 percent of theirs. Under our policy, by contrast, the (much lower) maintenance costs attributed to each type of truck would be fully covered by the taxes.

A comparison of the budgetary consequences of our policy and current policy is shown in table 3-13. To represent steady-state expenditures,

Table 3-11. *Change in Tax Burden, Cost, and Welfare per Truck*

	Urban			*Intercity*		
Item	*Initial*	*Final*	*Change*	*Initial*	*Final*	*Change*
Tax per ton-mile (cents)	0.49	0.71	0.22	0.21	0.15	−0.06
Truckers' average total cost per ton-mile, including taxes (cents)	5.65	6.10	0.45	4.80	4.76	−0.04
Welfare per truck (dollars per year)	...	...	−309	...	...	251
Tax as percentage of initial average total cost (percent)	...	...	3.90	...	...	−1.30

Source: Authors' calculations.

we present pavement costs consisting of the annualized value of resurfacing expenditures (maintenance, excluding disruption costs) and the cost of the paving material itself (capital, which excludes all expenditures on grading, right-of-way, bridges, medians, shoulders, and drainage). Revenues are those raised by our proposed marginal-cost taxes or by the current (1982) taxes that we assumed would be replaced by them (that is, almost all those current taxes classified as user charges).

It is useful to consider the relationship between these expenditures and revenues, because it provides a benchmark for considering the need for other sources of revenue. We will refer to an excess of revenues over expenditures as the pavement surplus, and to a shortfall as the pavement deficit. We do not mean to imply by this that such a surplus or deficit should be eliminated. It is simply a way of summarizing this portion of the highway budget, a way that is especially useful because it helps identify technological properties of economies or diseconomies of scale that play a major role in highway finance.

Table 3-13 shows that the current large pavement deficit is reduced greatly by our proposed policy. (The $6.3 billion reduction corresponds to the combined changes in investment costs and revenues, less that portion representing disruption costs to motorists, from table 3-7.) The reason why a large pavement deficit remains under optimal pricing is durability economies: durability can be increased with a less than proportional increase in cost. Because the ability of a pavement to withstand traffic increases far more than proportionally with its thickness, the marginal cost of a heavy axle on an optimally designed heavy-

Table 3-12. *Contribution to Allocable Maintenance Costs, by Truck Type*[a]
Percent

	Current pricing and investment		*Optimal pricing and investment*[b]	
Type	*Share of allocable maintenance costs*	*Tax revenue contribution to allocable maintenance costs*	*Share of allocable maintenance costs*	*Tax revenue contribution to allocable maintenance costs*
		Intercity		
SU2	45.19	9.81	19.07	19.38
SU3	0.74	1.31	1.33	1.35
TT4	0.43	0.11	0.21	0.21
TT5	0.19	0.09	0.33	0.33
CS3	2.11	0.63	1.70	1.71
CS4	0.76	0.33	0.72	0.72
CS5	43.02	13.51	60.32	60.38
CS6	6.34	2.83	14.60	14.62
DS5	1.18	0.42	1.68	1.68
DS6	0.04	0.01	0.04	0.05
Total	100.00	29.05	100.00	100.43
		Urban		
SU2	88.24	11.88	67.52	67.92
SU3	3.10	1.25	9.61	9.67
TT4	0.61	0.13	0.96	0.96
TT5	0.06	0.01	0.09	0.09
CS3	0.58	0.18	2.49	2.50
CS4	0.25	0.07	0.61	0.61
CS5	5.82	0.40	6.95	6.96
CS6	0.90	0.13	10.51	10.52
DS5	0.38	0.07	0.87	0.87
DS6	0.06	0.01	0.39	0.39
Total	100.00	14.13	100.00	100.49

Source: Authors' calculations.

a. Allocable maintenance costs are the sum, over all vehicles, of the marginal maintenance cost of each vehicle. The columns show the portion of this attributable to each vehicle type or the portion covered by revenues from each vehicle type.

b. Revenues occasionally exceed costs because they include fixed fees.

Table 3-13. *Pavement Budget under Current and Optimal Pricing and Investment*
Billions of dollars a year

Item	*Current pricing and investment (1982)*	*Optimal pricing and investment*	*Change*
Pavement costs			
Maintenance (excluding disruption costs)	10.484	2.311	−8.173
Capital	9.636	10.913	1.277
Total	20.120	13.224	−6.896
Pavement revenues	3.955	3.381	−0.574
Pavement deficit	16.165	9.843	−6.322
Pavement cost-recovery ratio	0.197	0.256	. . .

Source: Authors' calculations. See text for explanation of items included. Cost-recovery ratio is revenues divided by costs.

duty road is quite small. Efficiency-based prices therefore do not attempt to extract from each truck a user charge that would pay for the entire pavement structure.

One may or may not find the pavement deficit, as an absolute measure, a helpful concept. But its reduction is a real, practical phenomenon: it improves the overall balance of highway revenues and expenditures and, hence, of the overall government budget. Particularly at the federal level, where budget deficits seem intractable, such a change must be viewed as an important practical advantage to the proposed policy.

Sensitivity of Findings

In the appendix to this chapter we investigate the sensitivity of our findings to changes in the specification of the model and in certain assumptions such as the interest rate, resurfacing costs, and pavement lifetimes. We also examine the effects of introducing user costs or traffic growth into the analysis. Most of the avenues we explore suggest that our basic findings are robust and do not change fundamentally even when we make significant changes in our assumptions.

If our estimates of existing pavement lifetimes are too high (by, for example, 30 percent), the annual welfare gain from our policy would increase (in this case, to $11 billion). But even if we have underestimated pavement lifetimes by the same percentage, the annual welfare gain would still be large, roughly $6 billion. Thus the benefits from optimal investment are robust to major changes in our characterization of current road design.

Consideration of user costs increases both marginal-cost user charges and the net benefits of our policy. These higher charges would entail some redistribution away from the trucking industry (and its shippers and customers)—about $0.8 billion a year, all from urban operations, but still far less than what would occur from a marginal-cost pricing policy alone. If either the interest rate or the capital cost for pavement were half as high as we assume, the combined pricing and investment policy would entail more durability, lower user charges, and welfare gains to truckers and their customers around $1.0 billion to $1.3 billion a year.

It is important to bear in mind, however, that any losses to truckers are overstated, and gains understated, because we have not taken into account the many ways they can respond besides shifting to different truck types. As already indicated, truckers will also have strong incentives to adjust loads, vehicle dispatching practices, and fleet sizes. For example, consider a trucking firm that uses four SU2s (two-axle single-unit trucks) to transport four 33,000-pound loads ten times a week; in response to a marginal-cost tax, the firm might ship the same goods using three CS5s (five-axle tractor trailors) each transporting a 55,000-pound load eight times a week. When one adds the potential cost reductions from shifting to routes with thicker pavements (hence lower user charges) and from relocating terminals that require extensive travel over high-priced thin pavements, it becomes clear that truckers would benefit more or be harmed less than we estimate.[25]

In addition to logistical responses, there will also be incentives for manufacturers to design trucks that minimize pavement damage through improved suspension, axle placement, and so on. Technological im-

25. It would be desirable to model these responses but the problems in doing so are formidable. Consider the joint response of vehicle type shifting and load size adjustments. A valid model would specify consistent functional expressions for these choices that could be used to recover the trucking firm's underlying profit function. We were unable to identify expressions that would yield tractable choice models. In addition, the entire tax schedule, describing how taxes vary with load for a given vehicle, changes as a result of our policy, so we would have to model load as a function of the entire tax schedule; not only is that difficult in itself, but it would be hard to estimate from current data because, as shown by table 3-4, current taxes do not vary much with axle weight.

provements in vehicle design could substantially reduce the tax burden for the trucking industry under our proposal.

Conclusion

It is commonly believed that the solution to the deterioration of America's roads is greatly increased public outlays. While there is no doubt that the existing system requires immediate and costly renovation, public costs over the long run can be greatly reduced by combining more efficient pricing with better investment guidelines. The required change in pricing policy is drastic, shifting from reliance on fuel taxes and weight-graduated license fees to direct mileage charges steeply graduated with respect to axle loads. We have estimated that such a pricing schedule in conjunction with a modest increase in capital outlays—costing only $1.3 billion on an annualized basis—can reduce annualized maintenance costs $9.4 billion a year, of which about 87 percent shows up in state and local highway budgets. The net result is a large reduction in the need for other funding sources to cover highway expenditures, yet with no overall increase in the amount of user charges paid either by heavy vehicles or by automobiles. We estimate a net welfare gain to the nation of nearly $8 billion a year.

The benefits may be even greater because our calculations account neither for the secondary ramifications of productivity gains that may come about from cheaper and easier road travel, nor for the additional efficiency-enhancing responses that truckers may make to further lower their tax burdens.

Simply pumping more money into the road system may help in the short run, but will not yield these kinds of dividends. We suspect that such a policy, after another couple of decades, would produce a crisis similar to today's, but larger. To provide a more permanent solution, we need the combination of greater pavement durability and incentives to reduce loadings that our policy provides.

Appendix

In this appendix, we summarize the results of altering our assumptions about the real interest rate, capital cost for additional pavement thick-

Table 3-14. *Results of Sensitivity Analysis: Real Interest Rate and Capital Cost*
Change, relative to current practice, in billions of 1982 dollars except as noted

	Economic effects of optimal pricing and investment				
	Base case	*Alternative assumptions*			
Item	*(table 3-7)*	r = 3	r = 9	0.5 k_2	2 k_2
Investment costs					
Maintenance savings[a]	9.428	11.007	7.759	10.114	7.958
Capital savings	−1.276	−1.330	−0.497	−1.177	−0.132
Total savings	8.152	9.677	7.262	8.937	7.826
Trucking firms' and shippers' welfare					
Intercity operations	0.720	1.565	−0.257	1.730	−1.346
Urban operations	−0.586	−0.509	−0.810	−0.411	−1.077
Total	0.134	1.056	−1.067	1.319	−2.423
Revenues[b]	−0.574	−1.393	0.447	−1.618	1.551
Modal shifting					
Modal surplus	0.029	0.031	0.040	0.031	0.065
Rail profits	0.011	−0.061	0.095	−0.077	0.187
Total welfare	7.752	9.310	6.777	8.592	7.206
Change in loadings (percent)	−38.12	−35.03	−40.90	−34.06	−43.54

Source: Authors' calculations.
a. Includes disruption costs to motorists, which account for approximately 13 percent of maintenance savings.
b. Revenues include gains or losses from modal shifting.

ness, user costs, lifetimes of existing pavements, and growth in traffic loadings.

All the calculations that compare costs undertaken at different times make use of an interest rate, r. It is the real long-term interest rate, that is, the long-term interest rate less the expected rate of inflation. In the text we take this to be 6 percent (see table 3-1). Here we show the results for 3 percent and 9 percent (see table 3-14).

Optimal durability can be expected to depend upon the capital cost of increasing durability of a newly constructed highway. The key parameter is k_2, the cost of the paving material (delivered and put in place) required to increase road thickness by one unit. We measured this at the values shown in table 3-1 from Federal Highway Administration construction price data, and confirmed its reasonableness through discussions with a paving contractor. Here we show how our results would change if k_2 were 50 percent lower or 100 percent higher. The results of these

Table 3-15. *Coefficients of Equation 2-4, Chapter 2, Estimated Using Alternative Specifications*

	Flexible pavements			*Rigid pavements*	
Coefficient	*Small-Winston*[a]	*Specification A*[b]	*Specification B*[b]	*Small-Winston*[a]	*Specification C*[c]
ln A_0	12.062	12.168	15.700	13.505	15.077
A_1	7.761	8.546	5.102	5.041	4.192
A_2	3.652	4.072	3.160	3.241	3.017
A_3	3.238	3.530	2.748	2.270	2.068

Source: Authors' calculations.

a. Small-Winston columns are identical to those in table 2-1.

b. Specifications A and B use the coefficients shown for flexible pavements (see text for explanation). In the case of specification B, the coefficients do not have precisely the same meaning because the definitions of the independent variables were changed; the same change is made when the coefficients are used to calculate the results of this table.

c. Specification C is Model A of the paper by Kenneth A. Small and Feng Zhang, "A Reanalysis of the AASHO Road Test Data: Rigid Pavements" (Department of Economics and Institute of Transportation Studies, University of California at Irvine, December 1988).

variations, which have only a modest effect on net welfare, are summarized in table 3-14 using the same format used in table 3-7.

Maintenance costs depend crucially upon the equation describing how long a pavement can last before it must be resurfaced. We estimated two such equations, one for rigid and one for flexible pavements, as described in chapter 2. One can raise many questions about the best functional form for such equations and about the most statistically efficient way to use the available data. We have estimated several such variations, obtaining new parameters to replace A_0–A_3 in table 2-1. Specification A replaces the coefficients for flexible pavements with those obtained by limiting the sample to road sections with structural number at least 2.0. Specification B replaces the coefficients for flexible pavements by those estimated from a slightly altered version of equation 2-4 of chapter 2, in which $(D + 1)$ is replaced by (D) and $(L_1 + L_2)$ is replaced by (L_1), thereby eliminating a rather arbitrary mixing of terms with different physical units. Specification C replaces the coefficients for rigid pavements with those obtained when one uses all the data with $q \leq 3.0$ instead of just those data with $q = 2.5$, and estimates the parameter b simultaneously with parameters A_0–A_3 (see footnote 22 in chapter 2). The actual coefficients used for these three variants are shown in table 3-15. The welfare calculations, which reveal that our findings are barely affected, are shown in table 3-16.

For practical reasons, our primary model excludes user costs in the determination of optimal durability and marginal cost. In table 3-17 we

Table 3-16. *Results of Sensitivity Analysis: Alternative Equations Determining Pavement Life*
Change, relative to current practice, in billions of 1982 dollars except as noted

	Economic effects of optimal pricing and investment			
Item	*Base case (table 3-7)*	*Specification A*[a]	*Specification B*[a]	*Specification C*[b]
Investment costs				
Maintenance savings[c]	9.428	9.587	9.385	9.441
Capital savings	−1.276	−1.303	−1.250	−1.274
Total savings	8.152	8.284	8.135	8.167
Trucking firms' and shippers' welfare				
Intercity operations	0.720	0.890	0.340	0.675
Urban operations	−0.586	−0.554	−0.661	−0.585
Total	0.134	0.336	−0.321	0.090
Revenues[d]	−0.574	−0.749	−0.185	−0.536
Modal shifting				
Modal surplus	0.029	0.029	0.033	0.030
Rail profits	0.011	−0.003	0.045	0.016
Total welfare	7.752	7.897	7.707	7.767
Change in loadings (percent)	−38.12	−37.56	−39.36	−38.28

Source: Authors' calculations.
a. Specifications A and B use the coefficients shown in table 3-15 for flexible pavements and the Small-Winston coefficients in table 2-1 for rigid pavements.
b. Specification C uses the coefficients shown in table 3-15 for rigid pavements and the Small-Winston coefficients in table 2-1 for flexible pavements.
c. Includes disruption costs to motorists, which account for approximately 13 percent of maintenance savings.
d. Revenues include gains or losses from modal shifting.

show the results of including user costs, using our best estimate of them as derived in the appendix to chapter 2. We assume that the all-vehicle traffic growth rate g is 2 percent a year,[26] and that average user cost is 32.5 cents per vehicle-mile on a new road and 6.9 percent higher on a road just prior to resurfacing[27]—that is, in equation 2-12 of chapter 2,

26. This is the number chosen for the intermediate example used in American Association of State Highway and Transportation Officials, *Proposed AASHTO Guide for Design of Pavement Structures* (Washington, D.C.: AASHTO, July 1985), Appendix D, p. D-29, with the comment: "Past experience shows that [this growth rate] is not uncommon."

27. The initial user cost is estimated at a money cost of 20 cents, based upon FHWA, *Highway Performance Monitoring System Analytical Process,* vol. II: *Technical Manual* (Department of Transportation, March 1983), tables C-2 and E-2, plus a time cost of

$v_0 = 0.325$ and $v_1/v_0 = 1.069$. Applying equations 2-14 through 2-16 of chapter 2 and the parameters in table 3-1 implies, for example, that marginal user cost MC_u would be 0.8 cents per esal-mile on a flexible pavement with a lifetime of twenty-five years and a traffic mix containing ten passenger cars for every one-esal truck.

Table 3-17 presents the effects of optimal pricing and investment when user costs are included in this way. The annual welfare gain is now $10.4 billion, about one-third larger than in the base case considered in table 3-7. This difference is accounted for largely by a $3 billion annual reduction in user costs. Capital costs are nearly a billion dollars greater in this scenario because there are now greater benefits from increased durability; but revenues and maintenance savings are more than a billion dollars greater, so the government budget situation is improved even more than in the base case. However, there is now a loss of truckers' and shippers' welfare of about $0.8 billion because user charges are about 27 percent higher than in the base case. (Some of this, admittedly, would be offset by the lower user costs themselves, of which truckers might receive around 10 percent, based on their share of the traffic mix and their somewhat higher per-mile operating costs than automobiles.)

Of course, the investments and user charges undertaken in our base case would also yield user cost savings, and we can estimate those savings using the same formulas and parameters. The result is $1.8 billion a year. Adding that to the total welfare gain of the base case provides a fairer comparison of the two policies and shows that if user costs vary as our best estimate shows, we lose only $0.8 billion in total welfare by ignoring them when setting the pricing and investment rules.

12.5 cents, based upon a speed of 60 miles per hour, occupancy of 1.5, and value of time per occupant of 5 dollars an hour. The percentage increase in each of these due to pavement deterioration to a pavement serviceability of 2.5 was estimated in two ways, and the results averaged. One way was to take measurements from the graphs in W. D. O. Paterson, *Road Deterioration and Maintenance Effects: Models for Planning and Management* (Johns Hopkins University Press for the World Bank, 1988), figures 2.2 and 2.3, leading to a 6 percent increase in money cost and a 1.5 percent increase in time cost. The other way was to use FHWA estimates for money cost reported by José A. Gomez-Ibañez and Mary M. O'Keeffe, *The Benefits from Improved Investment Rules: A Case Study of the Interstate Highway System*, Report DOT/OST P-34/86/030 (Department of Transportation, July 1985), pp. C-15 to C-17 (yielding a 12.5 percent increase); and for time cost from FHWA, *Highway Performance Monitoring System*, p. IV-24 (yielding a 4.6 percent increase).

Table 3-17 *Results of Sensitivity Analysis: User Costs and Pavement Life under Current Design*
Change, relative to current practice, in billions of 1982 dollars except as noted

	Economic effects of optimal pricing and investment			
Item	*Base case (table 3-7)*	*User costs included*	*0.7 T^0*	*1.3 T^0*
Investment costs				
Maintenance savings[a]	9.428	9.990	13.326	7.315
Capital savings	−1.276	−2.159	−1.838	−0.785
Total savings	8.152	7.831	11.488	6.530
User cost savings	...[b]	3.026	...	...
Trucking firms' and shippers' welfare				
Intercity operations	0.720	0.026	0.720	0.720
Urban operations	−0.586	−0.814	−0.586	−0.586
Total	0.134	−0.788	0.134	0.134
Revenues[c]	−0.574	0.195	−0.574	−0.574
Modal shifting				
Modal surplus	0.029	0.037	0.029	0.029
Rail profits	0.011	0.076	0.011	0.011
Total welfare				
Excluding user cost savings	7.752	...	11.088	6.130
Including user cost savings	9.563	10.377	...	...
Change in loadings (percent)	−38.120	−40.427	−38.120	−38.120
Weighted average user charge (cents per esal-mile)				
Intercity operations	1.39	1.86	1.39	1.39
Urban operations	6.70	8.18	6.70	6.70

Source: Author's calculations.
a. Includes disruption costs to motorists, which account for approximately 13 percent of maintenance savings.
b. Using the same method of calculating user costs as in the second column, the investments and traffic-loadings reduction of the base case would produce user cost savings of $1.811 billion.
c. Revenues include gains or losses from modal shifting.

As explained earlier, because we lack accurate data on the designs of existing roads, we have chosen pavement lifetimes T^0 that we believe are representative of actual existing road design, then calculated the road thicknesses that, according to our pavement equations, would produce that lifetime given currently observed traffic. Our optimal durability calculations produce new road thicknesses and pavement lifetimes. These optimal pavement lifetimes are fairly robust to changes in our model, but the differences between current and optimal investment are sensitive to our depiction of current design. Hence, we calculate the results of decreasing or increasing the assumed initial lifetimes T^0 by 30 percent. These results, which show a larger welfare gain if lifetimes were overestimated and a fairly small reduction in the gain if lifetimes were underestimated, are also reported in table 3-17.

We also investigated the effects of assuming a rate of growth in traffic loadings of 4 percent a year. In these calculations, we assume that the observed level of traffic loadings represents the average over the first pavement cycle ($\overline{Q}$ in chapter 2 appendix). Hence the assumed pavement lifetimes T^0, which we do not change, imply approximately the same thickness of pavement as when no traffic growth was assumed. We calculated the results of optimal investment at current traffic levels, a scenario presented using our standard assumptions in table 3-3; the results (not shown in the table) are that maintenance savings are $0.57 billion less and additional capital costs are $0.45 billion more than when no traffic growth was assumed, so the total savings are $5.28 billion instead of $6.30 billion.

CHAPTER FOUR

Simplifying Policy Administration

OUR ANALYSIS of pricing and investment policies thus far has presumed that complex investment rules and marginal-cost tax schedules would be accurately implemented. Administrators at the Federal Highway Administration would have to calculate optimal durability and price for at least 160 combinations of functional class of road, type of pavement, and interval of traffic volume. A truck traveling one hundred miles could easily use ten differently priced pavement segments. Such a scheme is not impossible to manage, especially with the help of new technology, but both truckers and administrators are likely to quail at the prospect.

In this chapter we evaluate three possible ways to simplify administration of our policy recommendations. First, we consider whether it is necessary to adopt optimal pricing and optimal investment simultaneously. Could either policy alone produce substantial benefits in the road sector while simplifying administrative procedures? Second, we consider whether simplified pricing and investment rules could produce welfare gains similar to those produced by optimal pricing and investment. Finally, we consider an alternative policy involving weight limits.

Pricing vs. Investment

One way of approaching the question of whether it is necessary to implement both optimal pricing and optimal investment is to ask whether society would gain much from optimal pricing once roads are optimally built and whether society would gain much from optimal investment once roads are optimally priced.

As shown in table 4-1, a policy of optimal road-wear pricing under current investment yields a $5.4 billion annual welfare gain—roughly two-thirds of the gain under optimal pricing and investment. But that net gain conceals a $5.6 billion loss to the trucking industry and its customers, a loss that could be softened but not eliminated by logistics adjustments described in chapter 3. Such a loss clearly would arouse opposition from

Table 4-1. *Economic Effects of Optimal Pricing at Current Investment*
Change, relative to current practice, in billions of 1982 dollars

Item	*Effect*
Maintenance savings[a]	6.441
Trucking firms' and shippers' welfare	
Intercity operations	−3.697
Urban operations	−1.889
Total	−5.586
Revenues[b]	3.884
Modal shifting	
Modal surplus	0.204
Rail profits	0.411
Total welfare	5.354
Change in loadings (percent)	−48.38

Source: Authors' calculations.
a. Maintenance savings includes local roads, and about 13 percent of it is in disruption cost savings to motorists.
b. Revenues include gains or losses from modal shifting.

truckers and from shippers, to whom much of the cost would be passed on. To be sure, with current average charges per axle passage of 1.34 cents and marginal costs that range from 1.48 cents to 125.45 cents (table 3-3), truckers are hardly paying their fair share now. But there is also merit to the truckers' argument that their trucks would damage roads far less if roads were correctly built in the first place. We will return to this dilemma in our policy discussion in chapter 7.

The converse policy, optimal investment under current pricing, seems more promising initially. Table 3-3 showed the changes in durability that such a policy would bring about. Table 4-2 shows the economic effects. The welfare gain of $6.3 billion—80 percent of the optimal gain—does not require a major redistribution among freight transport participants.

This is certainly an important finding. It verifies, up to a point, the conventional wisdom that significant capital investment in U.S. roads would bring large benefits. But optimal investment alone provides no incentives for truckers to shift to less damaging vehicles and thus is $1.1 billion less effective in reducing the long-term pavement deficit than our base case policy of optimal investment and pricing combined. Furthermore, its short-term effects on the budget are considerably worse because the capital investment entailed in this policy is nearly twice that in our base-case policy. For budgeting purposes, the present value of increased

Table 4-2. *Economic Effects of Optimal Investment at Current Pricing*
Change, relative to current practice, in billions of 1982 dollars

Item	*Effect*
Investment costs	
Maintenance savings[a]	8.536
Capital savings	−2.236
Total savings	6.300
Change in pavement budget relative to current situation	
Pavement costs	
Maintenance (excluding disruption costs)	−7.426
Capital	2.236
Total	−5.190
Pavement revenues	0
Pavement deficit	−5.190

Source: Authors' calculations.
a. Maintenance savings includes local roads, and about 13 percent of it is in disruption cost savings to motorists.

capital costs must all be incurred when a road is built or upgraded, whereas the maintenance savings are not seen for ten years or more. It is not difficult to see what such a policy would mean for a nation already struggling with an intractable federal budget deficit. Furthermore, the political climate is now favorable for extracting some concessions from the trucking industry, and the restructuring of user charges does not, as we have shown, hurt the industry in the long run. We believe it would be a great mistake to neglect this opportunity to put user charges on a more rational system, thereby yielding significant future gains.

In short, either optimal investment alone or optimal road-wear pricing alone could generate substantial benefits and help eliminate administrative complexity; but political difficulty would be greater for either policy alone than for both combined.

Simple Pricing Rules

Table 4-3 presents the effects of two policies using pricing schedules much simpler than those we have used so far. One policy would involve a uniform tax per standard axle passage applied to all roads; the other would involve a two-part tax with one rate applied to freeways and

Table 4-3. *Economic Effects of Welfare-Maximizing Uniform and Two-Part Freeway-Nonfreeway Pricing at Optimal Investment*
Change, relative to current practice, in billions of 1982 dollars, except as noted

Item	*Optimal road-wear charge per standard axle passage*		
	Uniform tax[a]	*Two-part tax*[b]	*Marginal-cost (base case)*
Investment costs			
Maintenance savings[c]	9.306	9.359	9.428
Capital savings	−1.208	−1.217	−1.276
Total savings	8.098	8.142	8.152
Trucking firms' and shippers' welfare			
Intercity operations	−1.702	−0.751	0.720
Urban operations	0.148	−0.129	−0.586
Total	−1.554	−0.880	0.134
Revenues[d]	0.827	0.262	−0.574
Modal shifting			
Modal surplus	0.081	0.057	0.029
Rail profits	0.241	0.155	0.011
Total welfare	7.693	7.736	7.752
Change in loadings (percent)	−44.24	−42.28	−38.12

Source: Authors' calculations.
a. Uniform tax of 3 cents an esal-mile.
b. Two-part tax of 0.11 cents an esal-mile for freeways, 7.82 cents an esal-mile for nonfreeways.
c. Maintenance savings includes local roads, and about 13 percent of it is in disruption cost savings to motorists.
d. Revenues include gains or losses from modal shifting.

another to nonfreeways. In each case the tax rate or rates were set by a search procedure so as to maximize total welfare. The base-case policy is repeated from table 3-7 to facilitate comparison.

Surprisingly, the simpler taxes produce only slightly lower welfare gains—less than a 1 percent difference—than marginal-cost taxes.[1] The explanation appears to be that the size of total welfare improvement depends mainly on reducing loadings through shifts in vehicle types, and that the key to achieving this reduction is changing the basis of the tax from gross vehicle weight to loadings. Once that change has been made, the precise levels of the tax primarily affect distribution.

The simpler tax schemes do cause a moderate redistribution from the trucking industry to the public treasury, thus reducing their political attractiveness compared with our base-case policy. Furthermore, their

1. Even under current investment, optimal uniform pricing produces only 5 percent less improvement in welfare than marginal-cost pricing.

Table 4-4. *Economic Effects of Optimal Pricing and Investment with Route Shifting*
Change, relative to current practice, in billions of 1982 dollars, except as noted

Item	*Effect*	
	Marginal-cost pricing	*Uniform pricing*[a]
Investment costs		
Maintenance savings[b]	9.459	9.306
Capital savings	−1.048	−1.208
Total savings	8.411	8.098
Trucking firms' and shippers' welfare		
Intercity operations	0.749	−1.702
Urban operations	−0.580	0.148
Total	0.169	−1.554
Revenues[c]	−0.605	0.827
Modal shifting		
Modal surplus	0.030	0.081
Rail profits	0.021	0.241
Total welfare	8.026	7.693
Change in loadings (percent)	−38.09	−44.24

Source: Authors' calculations.
a. Uniform tax of 3.0 cents per esal-mile.
b. Maintenance savings includes local roads, and about 13 percent of it is in disruption cost savings to motorists.
c. Revenues include gains or losses from modal shifting.

welfare effects relative to our base case may appear more appealing than they really are because our calculations do not take into account truckers' logistics responses to the marginal-cost tax. Incorporating route shifting could reveal substantial differences between a tax that varies by road durability and one that does not. To explore this possibility, we developed a simple model of route shifting based on assumed cross-elasticities of demand for travel on three different classes of road. That model, described in the appendix, portrays how truckers would shift routes in response to taxes that are relatively higher for thinner pavements, traveling more on freeways and less on collectors. But capturing this response still did not lead to large differences in the welfare effects, which are listed in table 4-4: the net welfare improvement from marginal-cost taxes was still only about 4 percent higher than the improvement from the optimal uniform price.

We conclude that an optimal uniform road-wear charge per standard axle passage may yield benefits that are nearly as large as those produced by a more detailed marginal-cost pricing schedule. However, we cannot

Table 4-5. *Welfare Gain and Pavement Budget under Uniform Taxes and Optimal Investment*
Billions of 1982 dollars

Uniform tax (cents per esal-mile)	*Welfare gain*	*Pavement budget surplus*	*Truckers' and shippers' welfare*[a]
3.0	7.693	−8.496	−1.554
5.0	7.559	−5.830	−4.736
7.0	7.241	−3.545	−7.744
9.0	6.808	−1.511	−10.617
10.6	6.414	0.014	−12.835
11.0	6.309	0.382	−13.380
13.0	5.783	2.159	−16.055
15.0	5.248	3.851	−18.658
17.0	4.720	5.534	−21.203

Source: Authors' calculations.
a. Does not include modal shifting.

be sure without more sophisticated modeling, and the uniform charge appears to be less politically expedient. We would therefore urge that both schemes be more fully evaluated for administrative feasibility.

What impact would uniform road-wear taxes have on the pavement deficit? In table 4-5 we present the trade-off between welfare and the pavement budget for a range of uniform taxes. The optimal uniform tax of 3 cents an esal-mile yields a pavement deficit of \$8.50 billion (as against the \$9.84 billion deficit under marginal-cost pricing). The deficit is eliminated at a much higher uniform tax of 10.6 cents, at which level the welfare gain is \$1.3 billion less than at the optimal uniform tax. This apparently small sacrifice in welfare for fiscal health may be appealing to some policymakers, but the distributional impact of the "break-even" uniform tax would appear to be so large as to be politically unworkable. The trucking industry would stand to lose nearly \$13 billion. If eliminating the pavement deficit is a priority, it may be more feasible to accomplish it by combining marginal-cost pavement wear taxes with congestion taxes, a policy we investigate in chapter 6.

Simple Investment Rules

Can investment rules also be simplified? We wondered whether the benefits we describe are dependent upon unrealistic precision in design-

Table 4-6. *Sensitivity of Investment Results to Errors in Estimating Traffic Loadings*
Billions of 1982 dollars

Estimation of traffic loadings	*Reduction in annual maintenance and capital costs*
Loadings estimated correctly	6.300
Loadings consistently overestimated by 50 percent	6.109
Loadings consistently underestimated by 50 percent	5.729
Estimates of loadings contain uniformly distributed random errors ranging from 50 percent below to 50 percent above actual loadings	6.192

Source: Authors' calculations.

ing highways. To test this, we examine in table 4-6 the sensitivity of the combined maintenance and capital cost savings to errors in estimating traffic loadings. Even large errors do not affect these savings by more than 10 percent. Similarly, we noted earlier that accounting for traffic growth had little impact on results. Hence the bulk of the welfare gains that our models predict should be available in a more realistic setting of imperfect highway-design capabilities.

Weight Limits

Using weight limits to reduce loadings is an obvious alternative to marginal-cost taxation. A direct comparison cannot be made here because we are not able to model the choice of load size.[2] We can, however, illustrate some possible roles for weight limits, either as an independent policy tool or as a complement to optimal pricing and investment. Recall that our model of truck type choice took current weight limits, which are by no means optimal, as given; welfare might

2. A useful comparison would be one between marginal-cost taxes and welfare-maximizing weight limits. Unfortunately, because our model treats load as exogenous, increasing weight limits increases welfare without bound. If load were endogenous, increasing weight limits would lead to higher loads that would increase maintenance costs, thus giving rise to a welfare maximum.

Table 4-7. *Gross Weight Limits Based on Weight Limits per Axle*
Pounds

Truck type	*Current limit*	*Modified limit*
SU2	40,177	40,000
SU3	54,872	54,000
TT4	77,838	80,000
TT5	81,951	94,000
CS3	60,065	60,000
CS4	74,167	74,000
CS5	81,167	88,000
CS6	82,940	104,706
DS5	82,549	100,000
DS6	83,793	114,000

Source: Current limits are 1982 national averages from the American Trucking Association. Modified limits are estimated safety limits from Transportation Research Board, *State Laws and Regulations on Truck Size and Weight*, Report 198 (Washington, D.C.: National Research Council, February 1979). The limit for the CS6 and independent corroboration comes from Implementation Planning Subcommittee, "Recommended Regulatory Principles for Interprovincial Heavy Vehicle Weights and Dimensions" (Ottawa, Canada: Road and Transportation Association of Canada, June 1987).

therefore be further improved, even in the presence of marginal-cost pricing, by adjusting weight limits.

One policy might be to modify weight limits, setting as the limit the maximum gross vehicle weight for which our truck types could be operated safely. Table 4-7 compares one estimate of such safety standards, from a study of truck safety by the Transportation Research Board, with current (1982) weight limits. The weight limits for trucks with two or three axles would fall slightly, while the limits for trucks with four to six axles (except the CS4) would rise significantly. This schedule encourages truckers to shift in the same direction as that encouraged by optimal pricing—that is, to shift to vehicles with more axles, in order to take advantage of the increased flexibility associated with higher weight limits. Truckers' welfare will improve if the benefits of this flexibility outweigh the reduced appeal of smaller trucks.

Using our model, we can simulate the resulting type shifting because we included weight limits among the variables influencing type choice (both directly, and indirectly through their effect on the dummy variable for illegality of a given truck type with the stated load). The results are presented in table 4-8.

We find that the new weight limits effectively complement optimal pricing and investment: the welfare gain increases from $7.75 billion to

Table 4-8. *Economic Effects of Adjusted Weight Limits with Optimal Pricing and Investment and with Current Pricing and Investment*

Change, relative to current practice, in billions of 1982 dollars

	Effect	
Item	*With optimal pricing and investment*	*With current pricing and investment*
Investment costs		
Maintenance savings[a]	9.495	4.073
Capital savings	−0.982	0.000
Total savings	8.513	4.073
Trucking firms' and shippers' welfare		
Intercity operations	4.231	3.259
Urban operations	−1.097	−0.591
Total	3.134	2.668
Revenues[b]	−0.651	0.046
Modal shifting		
Modal surplus	0.044	0.026
Rail profits	0.026	0.076
Total welfare	11.066	6.889
Change in loadings (percent)	−49.48	−41.68

Source: Authors' calculations

a. Maintenance savings includes local roads, and about 13 percent of it is in disruption cost savings to motorists.

b. Revenues include gains or losses from modal shifting.

more than $11 billion. Intercity truckers' welfare is enhanced, whereas urban truckers suffer some additional welfare loss because they prefer the maneuverable single-unit trucks that are subjected to tighter limits. Because no changes in load are accounted for here, however, the quantitative results are not at all certain.

The new weight limits also appear to improve welfare substantially even under current pricing and investment. Although we are not very confident in this use of our model, it does suggest that changes in weight limits might reduce loadings as much as marginal-cost taxes do and might provide substantial benefits to truckers.[3]

3. This finding is consistent with the notion behind the Turner proposal for weight-limit reform, which is under study by national highway organizations (see chapter 2, note 2).

Conclusion

Optimal pricing and optimal investment should be implemented jointly, but both can be implemented in a simplified way. Welfare-maximizing freeway and nonfreeway prices, or even a uniform price, which would lessen administrative burdens, can be implemented instead of detailed marginal-cost prices without much loss in efficiency. Implementation of optimal investment policy is not particularly dependent upon forecasting traffic loadings with great precision. Changes in weight limits that reduce vehicle loadings, without harming truckers, could also enhance welfare. Enforcement of such limits would actually be aided by our pricing policy because it requires keeping close records of loads carried.

Appendix

This appendix describes a route-shifting model. Let demand for truck travel (ton-miles) over route i (where i is the choice of freeway, arterial, or collector) be specified as a function of the prices P_j of each of these three routes:

$$D_i = D_i(P_1, P_2, P_3).$$

The change in ton-miles for route i is approximately

$$\Delta D_i = \sum_{j=1}^{3} \frac{\partial D_i}{\partial P_j} \Delta P_j = \sum_{j=1}^{3} \left(\frac{P_j}{D_i} \frac{\partial D_i}{\partial P_j} \right) \frac{D_i}{P_j} \Delta P_j.$$

Hence,

$$\Delta D_i / D_i = \sum_{j=1}^{3} E_{ij} \left(\frac{\Delta P_j}{P_j} \right),$$

where E_{ij} is the elasticity of demand for route i with respect to the price of route j, and $\Delta P_j / P_j$ is the percentage change in average cost per ton-mile for route j. To implement this model we initially assume cross-elasticities of 0.5. Own elasticities are set at values, typically around −0.7, that keep total ton-miles constant. We also assume the change in initial ton-miles on any route cannot exceed 30 percent. Because taxes are a small part of average costs (see table 3-11), changes in the tax did not have a large impact on route shifting.

CHAPTER FIVE

Congestion and Highway Capacity

SCARCE CAPACITY has long posed frustrating and intractable road-pricing problems for policymakers. Providing sufficient capacity to accommodate peak traffic volumes is by far the most expensive part of the public responsibility for roads in the United States. No matter how many "solutions" to the problem of urban highway congestion—ride sharing, mass transit, higher parking fees—are tried, the problem grows worse. One approach that has not been tried in the United States is called congestion pricing. At the moment it is an idea resisted by most policymakers who know of it and wholly unknown to the vast majority of voters and politicians.[1] But we believe not only that congestion pricing will work but that the time may be right for a new political consensus incorporating it into a highway financing package.

The Current Congestion Problem

The American public and news media seem to need no convincing that urban highway congestion is a serious and growing problem. The late 1980s have been marked by an outpouring of news articles, columns, letters to editors, and television specials highlighting congestion. Surveys find highway congestion high on people's list of concerns, and severely deteriorating levels of service are commonplace.

Special studies in many metropolitan areas have documented certain instances of worsening congestion. Average weekday peak-hour delays crossing the Hudson River into Manhattan roughly doubled over the past ten years.[2] Speeds on many sections of the Capitol Beltway around Washington, D.C., fell anywhere from 15 to 50 percent in the past six

1. For example, congestion pricing is never mentioned in the eight-page cover story, "Gridlock! Congestion on America's Highways and Runways Takes a Grinding Toll" in *Time* (September 12, 1988), pp. 52–60.

2. Richard Levine, "Car Madness in Manhattan: Cure Sought," *New York Times*, October 11, 1987, p. E-6.

years.[3] Congestion delays in the San Francisco Bay Area grew more than 50 percent in just two years.[4] In the greater Los Angeles area, average peak-hour traffic at ten points on major arterials rose 45 percent between 1970 and 1980; total hours of traffic delay in the five-county area grew 30 percent between 1982 and 1984 and are projected to more than triple between 1984 and 2010.[5] Seattle's 1980 average commuter travel time of twenty-two minutes is expected to rise to thirty-four minutes by 1990.[6]

Nationwide data on congestion are too recent to document these trends for the nation as a whole. Past data on average speed are unreliable, limited to certain highways, and highly politicized because of certification requirements for the national speed-limit law. Census data on commuting times and distances show little change, partly because people change jobs and residences to maintain roughly constant commuting times in the face of changing conditions and partly because increased congestion is offset by the opening of new, remote, and initially uncrowded commercial and residential areas.

Related data, however, suggest that congestion is probably getting worse in most metropolitan areas, not only those just noted. Table 5-1 shows the proportion of peak-period traffic that uses highways with traffic volume exceeding 80 percent of design capacity, an indicator of congestion. That proportion rose steadily and dramatically on interstates, from 42 percent in 1975 to 63 percent in 1987, while for other arterials it remained about constant. Those figures probably understate the increase in overall exposure of the population to congestion, because in many areas the peak period has lengthened. (In fact, the Federal Highway Administrator says that traffic delays on noninterstate urban highways rose 30 percent in just one year, from 1984 to 1985.)[7]

3. John Lancaster, "Beltway Rush-Hour Commuters Find the Going Getting Slower," *Washington Post*, April 21, 1988, pp. D1–2.

4. California Assembly Office of Research, *California 2000: Gridlock in the Making* (Sacramento: CAOR, March 1988), p. 5.

5. Southern California Association of Governments, *1984 Regional Transportation Plan*, vol. 1: *Transportation Needs and Strategies* (Los Angeles: SCAG, 1984), p. II-15; California Assembly Office of Research, *California 2000*, p. 5; Southern California Association of Governments, *Impact Assessment: Draft Baseline Projection—Transportation* (Los Angeles: SCAG, January 1987), p. 9–21.

6. Robert Dunphy, "Urban Traffic Congestion: A National Crisis," *Urban Land*, vol. 44 (October 1985), pp. 2–7; item cited is on p. 3.

7. Statement attributed to Robert E. Farris, in Transportation Research Board, "In-

Table 5-1. *Percentage of Peak-Period Vehicle-Miles Traveled on Highways with Volume-to-Capacity Ratios Exceeding 0.80, Selected Years, 1975–87*

Year	*Interstate highways*	*Other arterials*
1975	42	42
1978	48	43
1981	49	43
1983	54	43
1985	61	40
1987	63	42

Sources: *The Status of the Nation's Highways: Conditions and Performance*, Committee Print, House Committee on Public Works and Transportation (GPO, various years). Figures updated by Federal Highway Administration.

Automobile ownership and use have also risen dramatically. Table 5-2 shows U.S. motor vehicle registrations between 1970 and 1986. In 1970, slightly more than one vehicle was registered for every 2 people; now the ratio is nearly one for every 1.4 people. Some of this increase is due to smaller households, but even the number of vehicles per household has risen significantly, from 1.7 to 2.0. Although some of this increase is offset by a decline in miles driven per vehicle, total vehicle-miles have still risen faster than either population or number of households; furthermore, the tendency toward two-car and three-car families has probably led to more of those miles being driven on short trips at congested times. Moreover, the rise in vehicle-miles traveled has been sharper in urban than in rural areas: the proportion of vehicle-miles in urban areas increased from 51.6 percent to 59.2 percent over the same period.

Road mileage, in contrast, has hardly changed at all. It increased only 4 percent during 1970–86, when vehicle-miles traveled grew 65 percent. Lane mileage probably did somewhat better, but data are unavailable.

The Federal Highway Administration has projected future congestion delays for a comprehensive sample of urban freeways. Total hours of delay are expected to quadruple, from 1.25 billion in 1984 to 6.9 billion in 2005, even with the most extensive series of road improvements believed feasible.[8]

Car Navigation System to Get Test Run in Los Angeles," *TR News*, May–June 1988, p. 28.

8. Calculated from Jeffrey A. Lindley, "Urban Freeway Congestion: Quantification

Table 5-2. *Motor Vehicle Registrations and Use and Road Mileage, Selected Years, 1970–86*

	Registrations			*Vehicle-miles traveled*			*Road mileage (thousands)*	
Year	*Total (millions)*	*Per person*	*Per household*	*Total (billions)*	*Per person*	*Per household*	*Total*	*Interstate*
1970	108.4	0.529	1.71	1,110	5,412	17,508	3,730	42.9
1975	132.9	0.616	1.87	1,328	6,148	18,673	3,838	43.0
1980	155.8	0.684	1.93	1,527	6,703	18,903	3,918	42.9
1983	163.7	0.697	1.95	1,653	7,040	19,697	3,880	43.0
1986	176.5	0.731	2.00	1,838	7,608	20,778	3,880	43.1

Sources: Registrations and vehicle-miles are computed from Bureau of the Census, *Statistical Abstract of the United States, 1987* (Department of Commerce, 1986), tables 991, 2, 56, and 1020; the figure for 1986 vehicle-miles traveled is from Federal Highway Administration, *Highway Statistics, 1986* (Department of Transportation, 1987), table VM-1, p. 177. Road mileage is given in FHA, *Highway Statistics*, for years in question: table M-21 for 1970 and 1975, table HM-18 for 1983 and 1986; because 1980 data were incomplete, the road mileages shown here for 1980 are actually those for 1979, from table M-21 in the 1979 volume.

As noted in chapter 1, revenues from highway user charges have not kept up with actual highway expenditures, much less with perceived needs for improvements. It is not surprising, then, to find that estimates of the expenditures needed for capacity expansion are far beyond current capabilities. For example, the Federal Highway Administration recently estimated the capital improvements required over a sixteen-year period to meet the "minimum standards considered acceptable from either a systems operation or safety standpoint."[9] Even omitting bridges, local roads, and roads where the improvements are physically infeasible, the projections called for expenditures in urban areas alone of $8.2 billion a year (1985 dollars) to increase capacity.[10] In fact, requirements for capacity increases needed merely to maintain 1983 performance levels

of the Problem and Effectiveness of Potential Solutions," *ITE Journal*, vol. 57 (January 1987), pp. 27–32, specifically from tables 1 and 6. An update of this study, using a more complete sample, forecasts a similar rate of increase in delay, starting with 3.11 billion vehicle hours in 1985: see Federal Highway Administration, "Urban and Suburban Highway Congestion," Working Paper 10 (Department of Transportation, December 1987), table 2-10, p. 24.

9. *The Status of the Nation's Highways: Conditions and Performance*, Committee Print, House Committee on Public Works and Transportation, 99 Cong., 1 sess. (GPO, 1985), pp. 58, 63–64.

10. *The Status of the Nation's Highways: Conditions and Performance*, Committee Print, House Committee on Public Works and Transportation, 98 Cong., 1 sess. (GPO, 1987), pp. 4–5, 57–62.

in urban areas are $5.8 billion a year—some $2 billion more than is spent on such improvements by federal and state governments, which provide virtually all funding for such capacity increases.[11]

Future technological advances might reduce these costs somewhat. In particular, automated sensing and controlling devices on highways and in cars could significantly increase the capacity of a highway lane. Such devices could also signal drivers to alert them to congestion on particular routes and suggest less crowded alternatives. Whether such "smart roads" and "smart cars" turn out to be cheaper than conventional methods for increasing capacity remains to be seen; but in any event their full-scale implementation remains far in the future, nor would it alter the fundamental causes of congestion during peak periods.

Funding levels under current approaches clearly fall far short of what would be required to reduce congestion or even keep it from becoming dramatically worse. The nation's urban areas are threatened with the collapse of one of their most crucial economic and human support systems. The only two alternatives are to find enormous new sources of funds or to find new ways to manage scarce highway capacity.

Reducing Congestion: Why the Experts Have Given Up

Numerous policies have been proposed to reduce urban highway congestion. Among them are ride sharing, operational improvements to road systems, higher parking fees, measures to encourage transit use, and better land-use planning.

Each of these policies can produce important benefits, as several comprehensive and thoughtful analyses show.[12] Nothing in our analysis

11. Amounts needed to maintain 1983 performances are from *Status* (1987), pp. 68, 71. The figures quoted are estimated capital requirements for the higher of the two traffic-growth scenarios shown (2.85 percent a year, the one preferred by the authors of the report), for the following categories only: reconstruction to freeway standards, reconstruction with added lanes, reconstruction with wider lanes, major and minor widening. Federal and state expenditures are calculated from *Status* (1987), pp. 11–12, 18, 20: urban disbursements of $9.774 billion (table II-6) are multiplied by the percentage spent on all categories excluding new construction and "RRR" (resurfacing, restoration, and rehabilitation). The latter is 50.7 percent (table II-8) assuming right-of-way and engineering expenditures are allocated across categories in the same proportions as other expenditures.

12. See especially John R. Meyer and José A. Gomez-Ibañez, *Autos, Transit, and*

refutes their value in improving urban life. But none of them will eliminate, or even substantially reduce, the severe congestion characterizing most of the nation's economically vital urban landscape. This view is more and more frequently espoused by transportation experts, whose pessimism is beginning to pervade the news media as well.[13]

The problem is that none of these policies accounts for the latent demand for peak-period highway travel. This latent demand consists of all potential peak-period highway users whose trips are now diverted or deterred by congestion itself. Any policy that makes some alternative to peak highway travel more attractive will founder on its own success, because any perceptible improvement in congestion will itself attract new peak-period highway users. These new users are people who now opt for some alternative that, were it not for congestion, would be perceived as inferior. Examples of such alternatives include carpooling in high-occupancy-vehicle lanes, using mass transit, living closer to one's workplace, traveling at less convenient times of day, traveling on a circuitous but less congested route, and doing without some trips altogether. Unless every alternative to peak highway travel is simultaneously made more attractive—a practical impossibility—the latent demand that exists today in virtually every major congested urban travel corridor will swamp the effects of a policy designed to entice people into more socially desirable behavior.

This principle, dubbed "the fundamental law of traffic congestion" by highway engineers, was articulated by Anthony Downs a quarter of a century ago for the case of expressways: "On urban commuter expressways, peak-hour traffic congestion rises to meet maximum capacity."[14] It is an example of a more general principle, known as the "tragedy of the commons," applying to any public good or environ-

---

Cities (Harvard University Press, 1981), especially chapter 11; Alan Altshuler with James P. Womack and John R. Pucher, *The Urban Transportation System: Politics and Policy Innovation* (MIT Press, 1979), especially chapter 9; and Kenneth A. Small, "Transportation and Urban Change," in Paul E. Peterson, ed., *The New Urban Reality* (Brookings, 1985), pp. 197–223, especially pp. 217–19.

13. For example, "The Traffic Explosion," *Washington Post*, June 12, 1988; Phil Yost, "Traffic Report: Backed up Forever," *San Jose Mercury News*, November 22, 1987; Jeffrey A. Perlman, "Experts Say There's Relief But No Cure in Traffic Initiative," *Los Angeles Times*, Orange County Edition, May 16, 1988.

14. Anthony Downs, "The Law of Peak-Hour Expressway Congestion," *Traffic Quarterly*, vol. 16 (July 1962), p. 393.

mental amenity that is available at little or no charge and whose quality deteriorates with intensity of use.[15]

The only way around the fundamental law of traffic congestion is to make peak-hour automobile driving less attractive through some means that does not entail significant social waste. One way to do that is to replace the existing deterrent, congestion itself, which involves enormous deadweight loss in the form of wasted time and frustration, with a monetary price that, aside from collection costs, represents simply a transfer of purchasing power with negligible loss of resources. Many users and potential users will then find the option of driving during peak hours as relatively unattractive as now—that is how they are induced to curtail their choices of that option. But because the transfer of purchasing power can be used for some other socially useful purpose, such as reducing license fees or fuel taxes, potential users are on balance far better off. Meanwhile, the urban highway system is restored as a functioning component of the system of urban production, making the entire urban economy run more smoothly and providing its members with higher incomes.

Chapter 2 made the same point in a more formal context. Sound pricing principles call for pricing the scarce capacity of highways to provide peak-period trips, just as we price other scarce goods in a modern economic system. This principle of peak-load pricing is identical to that used widely for telephones and increasingly for electricity, two private-sector industries where pricing by season and time of day has encouraged fuller use of expensive generating, transmitting, and switching capacity.

The Standard View of Congestion Pricing: Economic Utopia, Political Dystopia

Seldom has applied economics produced an idea with such unanimous professional conviction in both its validity and its political unacceptability. A. A. Walters's article on "congestion" in the *New Palgrave Dictionary of Economics* states flatly, "The best policy to deal with urban road congestion is likely to be some form of road pricing. However,

15. Garrett Hardin, "The Tragedy of the Commons," *Science*, vol. 162 (1968), pp. 1243–48.

road pricing is the exception rather than the rule."[16] The consensus among professional economists in favor of this approach on economic grounds is strong. The theory is now refined and standard; implementation has been widely explored; numerous empirical studies have predicted its effects; and the whole package has made its way into standard textbooks in urban and transportation economics.[17]

The reasons for the political unacceptability of congestion pricing are legion.[18] Capital for highways, unlike that for electricity and telephones, has usually been provided by the public sector; hence a different pricing ethic has prevailed. Travel is widely regarded as a basic right. People who mistrust governments may regard congestion pricing as a sinister form of tax increase. There is a widespread perception that implementing congestion pricing harms the poor, though this is not borne out by careful analysis and would in any event apply equally to electricity, telephones, or practically any private good.[19] Some implementation schemes involve

16. A. A. Walters, "Congestion," in John Eatwell, Murray Milgate, and Peter Newman, eds., *The New Palgrave: A Dictionary of Economics* (Macmillan, 1987), vol. 1, pp. 570–73. For a longer and excellent survey, see Steven A. Morrison, "A Survey of Road Pricing," *Transportation Research*, vol. 20A (March 1986), pp. 87–97.

17. For theory and empirical studies of effects, see Clifford Winston, "Conceptual Developments in the Economics of Transportation: An Interpretive Survey," *Journal of Economic Literature*, vol. 23 (March 1985), pp. 57–94; and Morrison, "A Survey of Road Pricing."

For implementation, see U.K. Ministry of Transport, *Road Pricing: The Economic and Technical Possibilities* (London: Her Majesty's Stationery Office, 1964); William Vickrey, "Pricing in Urban and Suburban Transport," *American Economic Review*, vol. 53 (May 1963, *Papers and Proceedings, 1962*), pp. 452–65; Ian Catling and Brian J. Harbord, "Electronic Road Pricing in Hong Kong, 2: The Technology," *Traffic Engineering and Control*, vol. 26 (December 1985), pp. 608–15; Kiran U. Bhatt, "What Can We Do About Urban Traffic Congestion?: A Pricing Approach," Urban Institute Paper 5032-03-1 (Washington, D.C.: Urban Institute, 1976).

For an example of treatment in a standard text, see Edwin S. Mills and Bruce W. Hamilton, *Urban Economics* (Scott, Foresman, 1984), chapter 12.

18. For insightful political analyses, see Thomas J. Higgins, "Road Pricing—Should and Might It Happen?" *Transportation*, vol. 8 (1979), pp. 99–113; Arnold M. Howitt, "Downtown Auto Restraint Policies: Adopting and Implementing Urban Transport Innovations," *Journal of Transport Economics and Policy*, vol. 14 (May 1980), pp. 155–67; Thomas J. Higgins, "Road-Pricing Attempts in the United States," *Transportation Research*, vol. 20A (March 1986), pp. 145–50; Ward Elliott, "Fumbling toward the Edge of History: California's Quest for a Road-Pricing Experiment," *Transportation Research*, vol. 20A (March 1986), pp. 151–56.

19. For an analysis of the effect of congestion pricing on the poor, see Kenneth A. Small, "The Incidence of Congestion Tolls on Urban Highways," *Journal of Urban Economics*, vol. 13 (January 1983), pp. 90–111.

electronic data collection that raises fears of invasion of privacy. People also fear adverse effects on business, in part because they fail to recognize the offsetting positive effects of the revenues from congestion pricing. Where toll roads are in place, the terms of bond financing may preclude a pricing system designed to curtail peak-hour traffic. Finally, federal capital incentives have encouraged overuse of the "build" solution to congestion.

At any rate, congestion pricing has been implemented in only one city in the modern world: Singapore. Since 1975, drivers wishing to enter the central business district of that city during the morning rush hour have been required to display on their cars a windshield sticker that is easily purchased and now costs the equivalent of approximately $2.50 a day. This scheme, extensively studied by the World Bank, "has been judged by both the government and independent observers as a considerable success on economic and political criteria."[20] Attempts to implement such a scheme in the United States, however, have never gotten beyond preliminary study, despite a serious federal attempt to stimulate pilot projects during the Carter administration.[21]

A Revisionist View of Congestion Pricing

Times change. Today strong new forces are at work that could give congestion pricing real popular appeal. Chief among them are the sheer desperation of drivers in metropolitan areas, technological advances, growing interest in toll roads, and fiscal realities.

An increasingly frustrated public routinely hears that conditions on highways are going to get worse and that no help is in sight. For example,

20. Walters, "Congestion," p. 573. See also World Bank, *Urban Transport* (Washington, D.C.: World Bank, 1985); Peter L. Watson and Edward P. Holland, "Relieving Traffic Congestion: The Singapore Area License Scheme," World Bank Staff Working Paper 281 (Washington, D.C.: World Bank, June 1978).

21. With grants from the U.S. Department of Transportation, the Urban Institute performed preliminary studies for Honolulu, Berkeley (California), and Madison (Wisconsin). In all three cases, political factors prevented further study or experimentation. See Higgins, "Road-Pricing Attempts," pp. 145–46; Melvyn D. Cheslow, "A Road Pricing and Transit Improvement Program in Berkeley, California: A Preliminary Analysis," Urban Institute Paper 5050-3-6 (Washington, D.C.: Urban Institute, 1978); Franklin Spielberg, "Transportation Improvements in Madison, Wisconsin: Preliminary Analysis of Pricing Programs for Roads and Parking in Conjunction with Transit Changes," Urban Institute Paper 5050-3-7 (Washington, D.C.: Urban Institute, 1978).

readers of the *San Jose Mercury News* were recently told that hours of delay due to congestion would rise 174 percent in California in just ten years—430 percent in Southern California by the year 2010—and that implementation of every project or mitigation strategy now contemplated would cut Southern California's increase only to 250 percent.[22] Anthony Downs's contribution to a collection of op-ed columns on congestion in the *Washington Post* is entitled "You'd Better Get Used to It."[23] Such pessimism coincides with unprecedented concern with congestion as a defect of urban life.

An important reason for this pessimism is a growing understanding of the "fundamental law of traffic congestion" noted earlier. The investment solution alone—that is, either widening or building roads—is not a solution to congestion, even though it would produce other benefits such as shortening the peak period and allowing more people to travel at more convenient times of day.

New technology will promote political acceptability for congestion pricing. Most people, when confronted with a suggestion for pricing roads, conjure up mental images of toll booths and long queues of cars. Yet even on conventional toll roads and bridges, toll booths are giving way to automated collection by a variety of new devices ranging from zebra-stripe stickers to tiny signal-activated radio transmitters. Collectively known as "automatic vehicle identification," these devices have been tested at sites as varied as the Lincoln Tunnel to New York City, the Aalesund Tunnel in Norway, the Coronado Bay Bridge in San Diego, and the city streets of Hong Kong.[24] They are also being considered for a number of newly proposed conventional toll roads in the United States and have been installed on some truck fleets in the western states as part

22. Yost, "Traffic Report."

23. "The Traffic Explosion," *Washington Post.*

24. Robert S. Foote, "Prospects for Non-Stop Toll Collection Using Automatic Vehicle Identification," *Traffic Quarterly*, vol. 35 (July 1981), pp. 445–60; Wiley D. Cunagin, *Use of Weigh-In-Motion Systems for Data Collection and Enforcement*, National Cooperative Highway Research Program, Synthesis of Highway Practice 124 (Washington, D.C.: Transportation Research Board, 1986); Science Applications, Inc., "AVI Study, Jacksonville Expressway System, Technical Memorandum 1: Technology Options" (La Jolla, California: SAI, undated); Ian Catling, "Automatic Vehicle Identification," in Peter Bonsall and Michael Bell, eds., *Information Technology Applications in Transport* (Utrecht, Netherlands: Science Press, 1986); Akos Szoboszlay, "Toll Collection Using Automatic Vehicle Identification," *California Transit*, no. 3 (Sacramento: California Transit League, April–June 1986), pp. 1, 6–9.

of a program that combines automatic weighing of trucks with logistical control by fleet owners.[25] Railroads, too, are using similar devices to help keep track of their cars' movements.[26]

The Hong Kong demonstration, involving eight to twelve months of operation by 2,600 government and volunteer vehicles, exceeded by a wide margin a set of stringent goals for reliability and economy. A sample of the positive conclusions from an evaluation study includes the following:

—more than 99.7 percent of vehicles crossing a given toll site were correctly identified;

—fewer than one in 10 million vehicles crossing a given toll site would be misidentified as the wrong vehicle in a full system;

—the electronic number plate attached to each vehicle would easily surpass its performance specification that 90 percent last twelve years without failure;

—security features could detect attempted fraud;

—automatic closed-circuit television cameras could reliably identify license numbers of vehicles not transmitting a valid identification (the cameras were designed not to photograph the driver);

—manual supplementary police enforcement at toll sites was proven feasible;

—the transmission, computer processing, and accounting systems worked almost flawlessly.

The evaluation concluded that "the overall performance is more than adequate for a full system, the system working reliably under a far wider range of conditions than expected."[27]

25. Loyd Henion and Barbara Koos, "The Heavy Vehicle Electronic License Plate System Development Program: A Progress Report," *Transportation Research Forum Proceedings*, vol. 27, no. 1 (1986), pp. 188–92; Peter Davies and Fraser K. Sommerville, "Development of Heavy-Vehicle Electronic License Plate Concept," *Transportation Research Record*, no. 1060 (1986), pp. 121–27; Dave Mathieu, "WIM and Electronic Truck Monitoring Gets Rave Reviews in State of Oregon," *Weighing and Measurement*, vol. 69 (August 1985), pp. 12–17.

26. John Armstrong, "Breakthroughs in Vehicle Identification," *Railway Age* (April 1984), pp. 40–48.

27. Catling and Harbord, "Electronic Road Pricing," p. 615. Sanford Borins also concludes that technological feasibility was demonstrated, though he derives from the Hong Kong experience a pessimistic conclusion about political feasibility: "Electronic Road Pricing: An Idea Whose Time May Never Come," *Transportation Research*, vol. 22A (January 1988), pp. 37–44.

The third factor in the changing political climate is a renewed interest in toll roads, arising mainly from the intensely felt need for highway improvements in growing metropolitan areas combined with a shortfall of funds from usual sources such as state gasoline taxes. New toll roads have been completed or are in planning stages in California, Colorado, Delaware, Georgia, Florida, Idaho, Oklahoma, Pennsylvania, South Carolina, Texas, and Virginia. Federal legislation known as the Surface Transportation and Uniform Relocation Assistance Act of 1987 permits for the first time the use of federal highway funds for up to 35 percent of the cost of building toll roads in selected areas.

There are several reasons to think that this trend will ultimately encourage congestion pricing. Some toll-road proposals already include the electronic identification technology that, once familiar, would remove a major practical barrier to peak-load pricing of roads. Also, funding shortfalls may elicit a closer look at the possibilities for using higher peak-period tolls to boost revenues.[28] Yet another reason stems from a likely side effect of toll roads in densely developed areas: the diversion of traffic to adjacent arterials and streets may stimulate interest in pricing those arterials and streets as a countermeasure. (Of course, this same phenomenon may also cause some resistance to toll roads in built-up areas.)

That the mundane concept of toll roads could lead to congestion pricing is illustrated by the "toll ring" around the central business district of Bergen, Norway. Since 1986, a toll has been charged on all entrance roads to the peninsular downtown area. The purpose is to raise money not for the entrance roads themselves, but for downtown street improvements. The tolls are charged only from 6 A.M. to 10 P.M. weekdays, making the system something of a hybrid between a conventional toll and a congestion price. Other cities in Norway are considering similar schemes.[29]

28. In August 1988, California State Senator John Seymour was induced by pessimistic forecasts of funding for three toll corridors in Orange County to seek, at some political cost, authorization for state contributions to construction. A study of one of these corridors suggests that more revenue could be raised by setting a higher toll rate during the peak period. See Austin-Foust Associates, with G. J. "Pete" Fielding and Kenneth A. Small, *San Joaquin Hills Transportation Corridor: Toll Road Feasibility Analysis* (Santa Ana, California: Austin-Foust Associates, December 1986), p. III-11.

29. Odd I. Larsen, "The Toll Ring in Bergen, Norway—The First Year of Operation," *Traffic Engineering and Control*, vol. 29 (April 1988), pp. 216–22.

Interest in solving problems of government finance also makes the political climate more favorable to congestion pricing. For government in general, the 1980s have brought new awareness of the harmful distorting effects of most taxes, making all the more attractive a revenue source that eliminates inefficiencies instead of aggravating them—especially in a time of persistent and widely feared federal budget deficits. At the same time, the financial health of large cities remains precarious, making a nondistorting revenue source even more attractive to a city government.

Already, congestion pricing has begun to receive favorable comment in the policy-setting arena. Several editorials and newspaper columns in California have recently supported the idea, which is included among the options suggested to the California state legislature by its legislative analyst and which has been publicly endorsed by Keith McKean, Director of the California Department of Transportation, District 12.[30] New York City's Transportation Commissioner reportedly advocates tolls in a portion of the city's central business district.[31] The draft air quality management plan for Southern California now includes congestion pricing as a "future study issue."[32] The Congressional Budget Office offers it as an example of a more general "concern from a managerial perspective [that infrastructure prices] encourage efficient use."[33] The California Assembly Office of Research argues that "California's current transportation dilemma stems from a mismatch of value and what is charged for that value," and offers "mileage-based fees" and a Singapore-type "permit system" as two ways to remedy this mismatch.[34] Internationally, serious discussion of congestion pricing is under way in

30. For the recent editorial support, see "Winning the War against Traffic," *Oakland Tribune*, March 2, 1988, p. A-10 (editorial); Dennis J. Aigner, "How to End Freeway Jams? Bill Drivers for Road Use," *Los Angeles Herald Examiner*, October 1, 1987, p. A15. For the recommendation of the legislative analyst, see *The 1988–89 Budget: Perspectives and Issues,* Report of the Legislative Analyst to the Joint Legislative Budget Committee (Sacramento: California Legislature, 1988), p. 202.

31. Kirk Johnson, "Unlocking the Gridlock for Those Who Must Drive," *New York Times*, July 17, 1988, p. E-7.

32. Southern California Association of Governments, *Draft Air Quality Management Plan: 1988 Revision* (Los Angeles: SCAG, May 1988), pp. 179–81.

33. Congressional Budget Office, *Federal Policies for Infrastructure Management* (GPO, 1986), p. 64.

34. California Assembly Office of Research, *California 2000*, pp. 47–48.

Stockholm, Sweden.[35] The *Economist*, Britain's premier business magazine, has come out strongly in favor of congestion pricing.[36]

The analysis of highway finance in this book offers yet another reason to consider congestion pricing. As we have shown, providing more durable roads would lower the overall long-run cost of providing for heavy vehicles, but efficient pricing of heavy vehicles would fail to recover the entire public cost even of the pavement, much less of the entire highway. The usual response to such a discrepancy is to establish other taxes or fees, usually introducing some economic distortions. As we show in the next chapter, adopting congestion pricing in conjunction with road-wear pricing not only eliminates existing distortions, but brings the entire highway budget very close to balance over a long period.

We do not expect congestion pricing to appear suddenly on every political candidate's campaign platform. Rather, we believe that the time may be ripe for a few small-scale experiments, perhaps of the kind originally planned by the Department of Transportation during the 1970s. Favorable results with them could instigate growing interest in larger demonstrations. Federal policy could greatly encourage such a trend by requiring demonstrations of innovative demand-side policies, aimed at minimizing the need for additional highway capacity, as a condition for funding new construction.

The Impact of Congestion Pricing

Congestion is different from road wear in one important respect: it occurs only at certain places and times. Furthermore, the cost of expanding capacity varies enormously from one place to another, often depending critically on the locations of buildings, other roads, interchanges, bodies of water, and other local features. Because the precise timing and level of congestion charges would be very specific to local conditions, a comprehensive nationwide analysis like that we have done for road wear is impossible. A rough idea of the likely effects of congestion

35. "Traffic Jams: The City, the Commuter, and the Car," *Economist* (18–24 February 1989), pp. 19–22.

36. "Make Them Pay," *Economist* (18–24 February 1989), pp. 11–12.

pricing, however, is available from studies done under various assumed conditions.[37]

Empirical studies of congestion pricing in major British and North American cities find that efficient peak-period tolls, at price levels of roughly a decade ago, would usually be in the range of 5 cents to 30 cents a mile per auto, or a dollar or two a day for typical commutes. The resulting reduction in peak traffic volume varies from case to case; estimates of 10 percent to 25 percent are common. For example, Philip Viton estimated that an optimal peak-period toll for the San Francisco-Oakland Bay Bridge would be $1.37 each way and would reduce auto traffic by 11 percent.[38] Kenneth Small, analyzing a simulated freeway corridor in the San Francisco Bay Area, estimated peak traffic reductions from congestion tolls at 10 percent for initial conditions of light congestion and 25 percent for initial conditions of heavy congestion.[39] Analysis of a proposed new tollway corridor in Orange County, California, suggested that a two-tier toll structure of 85 cents peak and 50 cents off-peak would reduce peak traffic 19 percent in 1995 and 30 percent in 2015 compared with reductions caused by a flat toll of 60 cents, even though the two toll structures would have virtually identical impacts on total daily traffic.[40] Projections of peak traffic reductions in Hong Kong, from a toll equivalent to approximately $2.00 a day for a typical commute, were 20 percent for a finely varied toll structure and 9 percent for a simpler scheme.[41] The Singapore toll of approximately $2.50 a day, which appears to have been initially far higher than optimal, was followed by a 69 percent reduction in peak traffic over four years.[42]

37. Morrison, "A Survey," pp. 88–93; also Mohring, "Relation between Optimum Congestion Tolls and Present Highway User Charges," *Transportation Research Record*, no. 47 (1964), pp. 1–4.

38. Philip A. Viton, "Equilibrium Short-Run Marginal Cost Pricing of a Transport Facility: The Case of the San Francisco Bay Bridge," *Journal of Transport Economics and Policy*, vol. 14 (May 1980), pp. 185–203.

39. Computed from Kenneth A. Small, "Bus Priority and Congestion Pricing on Urban Expressways," *Research in Transportation Economics*, vol. 1 (1983), pp. 27–74, especially pp. 32–33, 57–58.

40. Computed from Austin-Foust Associates and others, *San Joaquin Hills Transportation Corridor*, pp. III-8, III-11, plus data obtained from the study's authors.

41. Bill Harrison, "Electronic Road Pricing in Hong Kong, 3: Estimating and Evaluating the Effects," *Traffic Engineering and Control*, vol. 27 (January 1986), table II, p. 16.

42. Chee-Meow Seah, "Mass Mobility and Accessibility: Transport Planning and

Estimates of net social benefits for an entire large city are on the order of $20 million to $60 million a year. One study estimated that adopting congestion pricing throughout the United States would yield revenues of $54 billion a year (1981 dollars), which, after subtracting the direct welfare losses to road users, leaves net benefits of $5.65 billion a year—mostly in the form of annual travel-delay savings of approximately one billion vehicle-hours.[43] Of course, if congestion pricing were accompanied by increased investment in road capacity, congestion charges and road user welfare losses would be lower and net benefits probably even greater.

These benefits do not take into account the possibility that congestion pricing would greatly increase the use of public transit. For example, Philip Viton found that under one set of plausible conditions similar to those of the early 1970s, congestion pricing in San Francisco Bay Area highways would raise mass transit's share of downtown commuting trips to between 73 and 83 percent, compared with the actual 1970 share of 63 percent.[44] Depending on the response of mass transit agencies, this share increase could result in some combination of more efficient use of transit vehicles, more frequent service, and extension of service to new areas, all of which produce additional benefits.

One frequent objection to congestion pricing is that it would hurt the poor. Would it in fact be a regressive tax policy? Kenneth Small has examined this question by tracing the distributional impact of congestion pricing on a prototype urban expressway.[45] For an intermediate case in

---

Traffic Management in Singapore," *Transport Policy and Decision Making*, vol. 1, no. 1 (1980), pp. 55–71 (data cited are on p. 60).

43. FHWA, *Final Report on the Federal Highway Cost Allocation Study* (Department of Transportation, 1982), Appendix E, table 14; we have used the fact that the net benefits "are primarily travel time savings" (p. E-61) and that travel time is valued at $4.80 per vehicle-hour (p. E-31) to derive the travel savings as somewhat less than $5.65 billion divided by $4.80 per vehicle-hour. The FHWA study is also published as Douglas B. Lee, "New Benefits from Efficient Highway User Charges," *Transportation Research Record*, no. 858 (1982), pp. 14–20.

44. Philip A. Viton, "Pareto-Optimal Urban Transportation Equilibria," in *Research in Transportation Economics*, vol. 1 (1983), pp. 75–130; see p. 95 for the numbers cited.

45. Kenneth A. Small, "The Incidence of Congestion Tolls on Urban Highways," *Journal of Urban Economics*, vol. 13 (January 1983), pp. 90–111. For a similar study, but without the explicit calculation of effects of uses of revenues, see Yuval Cohen, "Commuter Welfare under Peak-Period Congestion Tolls: Who Gains and Who Loses?" *International Journal of Transport Economics*, vol. 14 (October 1987), pp. 239–66.

which initial congestion delay is six minutes a trip each way, the marginal-cost congestion toll is found to be nearly $1.00 for each round trip. Commuters are then divided into three income classes to trace the effects of their toll on each class. In the simplest approach, where the toll is considered in isolation, its incidence is mainly related to the proportion of commuters in each income class who drive to work. But this proportion rises less than proportionately with income: the proportion of commuters earning $30,000 a year who drive to work is not twice that of commuters who earn $15,000. Hence, viewed simply as a tax, the policy is regressive. Taking into account the value of time, assumed proportional to the commuter's after-tax wage rate, the impact is even more disparate, ranging from a 28 cent daily loss for the lowest income class to an 8 cent gain for the top.

The picture changes, however, when the uses of the revenues are taken into account. The effect depends, of course, on whether the revenues are used to reduce other taxes, to subsidize transit, or in some other way; hence, Small computed the effect under three quite different assumptions about how the revenue uses benefit each income class. Under all three assumptions, each income class gains from the package. For example, if the revenue uses provide equal value to each person, the resulting net benefit per commuter is 22 cents for the lowest income class and 86 cents for the highest; and even this disparity is due mainly to the higher value that the rich place on their time savings. Hence, it is misleading to characterize a congestion policy as "regressive" based simply on a tally of who pays the tolls.

To further understand the distributional impact of congestion pricing, it is helpful to examine the composition of the aggregate annual welfare gain of $135 per commuter (in 1972 prices) that arises from the case just described. First, consider calculations for all income classes combined. Those 67 percent of commuters who continue to use auto pay $250 a year in tolls, whereas bus riders pay none; so average toll payments are $168 per commuter. In addition, those 7 percent of commuters who switched to bus are inconvenienced by an average of $128 each, in the sense that they would be willing to pay this much a year to be exempt from the toll; this is an overall average loss of $9 a year per commuter. But the time saving of twelve minutes a day for both auto and bus commuters is valued on average at $144 a year. So the average commuter in his role as consumer pays $168 in tolls, suffers $9 in loss of convenience, and gains $144 in time savings, for a net loss of $33. But when benefits from use of the toll revenues are taken into account, even assuming they are worth no more than the amount of the revenues, the average consumer has a net gain of $135 a year.

Similar calculations for the lowest income class show that its average commuter pays $128 in tolls, suffers $17 in loss of convenience, and gains $73 in time savings for a net loss of $72; but if revenues are used to reduce taxes on an equal per-capita basis, his net effect is a $96 gain. This gain is composed of a positive money transfer of $40 plus $73 in time savings, offset by the $17 loss in convenience.

The upshot is that congestion pricing, coupled with an explicit plan for using the revenues, can benefit all income classes. By contrast, most taxes have undesirable side effects leading to a net loss, known as deadweight loss, even after revenue uses are accounted for. Because congestion pricing corrects rather than introduces economic distortions, it is inextricably tied to service improvement; assessing it through standard concepts of tax incidence is therefore misleading.

The burdens of a new tax or user charge are shifted throughout the economy through price adjustments. If congestion pricing were adopted, land values and wages would change as various competitive forces worked themselves out, altering the distribution of burdens and benefits from that calculated above.[46] Owners of urban land are particularly likely to be adversely affected, and this would shift at least part of the burden from road users to landowners, making it even more doubtful that low-income workers will be hurt. Congestion pricing may indeed make some individuals ultimately worse off, but these individuals should not be allowed to disguise their own self-interest by claiming that their losses characterize the effect of congestion pricing on the poor.

In practice, one likely use for the revenues from congestion pricing would be to reduce or eliminate the regressive vehicle registration fees and fuel taxes, now the largest components of highway user charges. The calculations in the next chapter suggest that the revenues in urban areas would be large enough to cover at least 80 percent of the long-run public costs of providing highways. Currently, as noted in chapter 1, only about 62 percent of public expenditures on highways are covered by any type of user charges. Therefore governments could use conges-

46. See J. Hayden Boyd, "Benefits and Costs of Urban Transportation: He Who Is Inelastic Receiveth and Other Parables," *Transportation Research Forum Proceedings*, vol. 17 (1976), pp. 290–97.

tion-fee revenues to eliminate all existing user taxes on cars and still have a great deal left over to reduce the subsidy from local property and sales taxes. Such a strategy would partially or fully compensate most people paying the congestion charges, thereby further broadening the range of people who are made better off by the entire package of policies.

Conclusion

Several trends are converging to make congestion pricing, long advocated unsuccessfully by economists, a serious possibility for curbing congestion and reducing the need for expensive highway capacity expansion. As we will show in detail in the next chapter, congestion pricing and road-wear pricing can be combined into a highway financing package that is nearly in long-run balance. It is impossible to provide good quantitative forecasts of the effects of congestion pricing if adopted nationwide, but studies to date suggest that tolls on the order of $1.00 to $2.00 per round trip for typical congested commutes might reduce round-trip travel time by ten to fifteen minutes per commuter, raise revenues of tens of billions of dollars annually, and provide some $5 billion in net benefits a year to society.

Congestion pricing is technologically feasible and economically well understood. It is the only urban transportation policy with a chance of substantially reducing congestion at the busiest times. It would also reduce the capital requirements and environmental impacts associated with highways by limiting their size. For all these reasons it is an important component of a fiscally sound policy of highway management.

CHAPTER SIX

Effects on Highway Finance

OUR ANALYSIS of road wear and congestion has emphasized an optimal policy of investment and pricing for both durability and capacity. Because we have treated road wear and congestion separately, the view of highway finance that we have presented has been necessarily incomplete. Focusing on road wear, we have seen that our recommended policy would leave the "pavement budget" in deficit—that is, the cost of the pavement itself and of its maintenance would not be covered by marginal-cost road-wear user charges. Focusing on congestion, we have seen that the costs of providing adequate capacity to meet expected traffic growth are far beyond the means of current funding programs, unless that growth is curtailed by congestion pricing.

But roads do not provide only for traffic volume or traffic loadings—that is, only for cars or for trucks. They provide simultaneously for both. A full picture of road costs and finance thus requires an analytical framework that takes both into account.[1] In the analysis that follows, we simplify our model of road wear and durability and combine it with an explicit model of congestion and road capacity, sticking as closely as possible to the formulations already introduced.[2] What we find is that combining our recommendations for policy on road wear and congestion leads to an urban road budget that is nearly in long-run balance.

1. The natural framework is the analytics of multiproducts or joint production, in which the relationship between pricing and cost coverage is explicitly addressed. This field of economics has blossomed in the last two decades and has already found important applications in other transportation services, particularly surface freight. For an excellent review, see Elizabeth E. Bailey and Ann F. Friedlaender, "Market Structure and Multiproduct Industries," *Journal of Economic Literature,* vol. 20 (September 1982), pp. 1024–48. For a discussion of the evolution in transportation analysis from simple single-product costing to sophisticated multiproduct cost models, see Clifford Winston, "Conceptual Developments in the Economics of Transportation: An Interpretive Survey," *Journal of Economic Literature,* vol. 23 (March 1985), pp. 57–94.

2. Because comprehensive data on road capacities and hourly traffic volumes are not available, we limit ourselves to analyzing two typical urban road segments rather than attempting to inventory the entire U.S. road system. Also, because we lack the ability to model accurately the demand for travel by location and time of day, we limit our analysis to cost functions, taking vehicular demands as given.

Returns to Scale and Budget Balance

The key to analyzing the road budget is to keep in mind the distinction between the production of a single product and the joint production of several products. In the production of a single product, the relationship between the revenues from marginal-cost pricing and the total costs of production is simple. Under constant returns to scale, they are equal; under increasing returns, where marginal cost is less than average cost, revenues fall short of costs; under decreasing returns, revenues exceed costs.[3]

If traffic loadings and traffic volume were handled on separate facilities, we could then analyze the budget balance of each facility under marginal-cost pricing by determining the type of returns to scale that apply to each. We have already done this for traffic loadings in chapter 3: since durability is supplied under sharply increasing returns to scale, the pavement budget defined there would be in deficit under optimal pricing and investment.[4] Although our proposed road-wear policy would reduce that deficit by $6.3 billion a year compared with current policies, it would still be nearly $10 billion annually.

A comparable analysis for traffic volume would require knowing whether the cost of providing highway capacity increases more or less than proportionally as capacity is increased. Here the evidence is less certain. Theoretical arguments can be provided on both sides. On the one hand, a single road of a given type probably has increasing returns for two reasons: first, capacity goes up faster than number of lanes (for example, two lanes in a given direction have more than twice the capacity of one lane); second, cost rises more slowly than number of lanes because certain features such as shoulders and median strip need not be expanded. On the other hand, a *system* of roads may have decreasing returns to scale because of the need for intersections, whose costs may well grow

3. These statements of standard textbook economics follow simply from the definition of economies of scale as the ratio of average to marginal cost. See Bailey and Friedlaender, "Multiproduct Industries," p. 1028. As they point out, that ratio is the reciprocal of the elasticity of cost with respect to output. Returns to scale are said to be increasing, constant, or decreasing if the ratio is greater than, equal to, or less than one.

4. This is a consequence of the fact that parameter A_1 in the equation describing the effect of road durability on road lifetime is much greater than one: see chapter 2, equation 2-4 and table 2-1.

more than proportionally to the widths of the roads (for example, the area of an intersection between two roads of equal width is proportional to the square of that width, so doubling the number of lanes quadruples the land and pavement required for intersections). Furthermore, in dense urban areas the road system occupies such a large fraction of available land that expanding the road may drive up the cost of land noticeably, which would tend to produce a rising average cost function—that is, decreasing returns to scale.[5]

These arguments, except for the rising land values, are reviewed in two studies that also provide empirical evidence and are among the few that carefully separate the effects of greater road capacity from the confounding effects of greater urbanization of the surrounding area. Theodore Keeler and Kenneth Small, using statistical analysis of data giving the actual construction costs for a variety of roads, find evidence of either constant or mildly increasing returns to scale; the best-fitting equation gives a ratio of average to marginal cost of 1.03, which is statistically indistinguishable from 1.00 (constant returns).[6] Marvin Kraus, using engineering specifications to estimate the costs of each component of a simple highway network, finds moderately increasing returns, with a ratio of 1.19 his best estimate.[7] Since neither of these studies includes the possibility of a rising supply price of land, they may overstate the degree of increasing returns in urban areas. We believe the most reliable assumption for urban areas is the estimate of 1.03 by Keeler and Small; later we discuss what happens if the ratio is 1.00 or 1.19.

Because traffic loadings and traffic volume are handled jointly on a single capital facility, not on separate facilities, the balance between total revenues and total costs in their production depends not just on

5. Returns to scale are sometimes defined as a property of a production function, in which case the definition we give here is a derived property of the corresponding cost function when factor prices are fixed. In that case, one should say that a rising supply price of land tends to cause "rising average cost" rather than "decreasing returns to scale." We, like Bailey and Friedlaender, ignore the distinction.

6. Theodore E. Keeler and Kenneth A. Small, "Optimal Peak-Load Pricing, Investment, and Service Levels on Urban Expressways," *Journal of Political Economy*, vol. 85 (February 1977), pp. 1–25. See especially pp. 5–10; as the text makes clear, there should be a minus sign on each of the estimates of parameter a_6 in table 1, p. 8.

7. Marvin Kraus, "Scale Economies Analysis for Urban Highway Networks," *Journal of Urban Economics*, vol. 9 (January 1981), pp. 1–22. See especially p. 20; his definition of "returns to scale" is the reciprocal of our definition of "economies of scale."

returns to scale for each, but on a second feature of production, known as economies of scope, which measures the extent to which producing a set of products jointly costs more (diseconomies of scope) or less (economies of scope) than producing them separately. In the case of highways, producing loadings and volume jointly costs more than producing each separately. The wider the road is made in order to accommodate more cars, the greater the cost of any additional thickness required to handle a heavy vehicle, because all the lanes must be built to the same thickness.[8] These diseconomies of scope counteract the scale economies in the separate production of loadings and volume and the resulting deficits under marginal-cost pricing. The reason is that once marginal-cost pricing is applied jointly to both cars and trucks, the cost of the pavement itself is charged for twice: once from trucks because they require a thicker pavement and once from cars because they require a wider pavement. The result of these offsetting tendencies turns out to be a budget approaching long-run balance.

A Model of Multiproduct Returns to Scale

We begin by defining two specific "products" that are created with the help of a road. One is traffic volume, V, measured as the number of passenger car equivalents that pass over the road during peak periods over an entire year. The other is traffic loadings, Q, measured as the number of equivalent standard axle loads that pass over the road during the year. As explained in chapter 2, any given vehicle contributes to both. But because the aggregate contribution of trucks to congestion (in the absence of accidents, which we do not include here) is small and the contribution of cars to road wear is negligible, for simplicity we may think of traffic volume as caused by cars and loadings as caused by trucks, in which case the two products are interpreted as peak auto travel (requiring capacity) and all truck travel (requiring durability).

8. H. Youn Kim makes an analogous argument for diseconomies of scope in the joint provision of passenger and freight rail services: "Keeping tracks smooth enough for passenger trains but heavy enough to withstand the axle loadings of freight trains can indeed be costly . . . [causing] diseconomies of scope." H. Youn Kim, "Economies of Scale and Scope in Multiproduct Firms: Evidence from U.S. Railroads," *Applied Economics*, vol. 19 (June 1987), p. 739.

The short-run cost of producing these products on a highway with W lanes and pavement thickness D is

$$\text{(6-1)} \quad SRTC(V,Q; W,D) = V \cdot c(V,W) + r \cdot M(Q,W,D) + r \cdot K(W,D),$$

where c is the average congestion cost (in dollars per passenger-car-equivalent–mile), r is the interest rate (expressed as a fraction), M is the present discounted value of all resurfacing expenses, and K is the cost of building the highway. As in earlier chapters, we exclude those costs to users that are invariant with congestion, because they are constant in this analysis and do not enter the public budget. (This formulation also assumes that the monetary value of a vehicle's time delay is proportional to its number of passenger car equivalents, an unimportant simplification given the low proportion of trucks in urban vehicle fleets.)

This is a slightly simpler version of the model explained in the appendix to chapter 2. Although W is called "number of lanes," it is really a continuous variable equal to capacity divided by c_w, the capacity of a "standard" lane; capacity can be altered by many means besides adding an entire lane, but it is simpler to relate the results to actual highways by using W as our unit of width. For example, $W = 2.2$ would describe two lanes with some improvements to shoulders, median, or intersections that effectively add 10 percent to their capacity.

Maintenance and capital costs, M and K, are specified consistently with the specification in chapter 2. Maintenance cost M decreases with the time between resurfacings, which depends upon D and Q; it includes disruption costs to motorists. Capital cost K is linear in width with a term proportional to D to reflect the cost of the pavement itself:

$$\text{(6-2)} \qquad K(W,D) = \begin{cases} 0 & \text{if } W = 0 \\ k_0 + (k_1 + k_2 D)W & \text{if } W > 0. \end{cases}$$

For congestion cost, we adopt a simple nonlinear form used by many previous authors and apparently first proposed by William Vickrey, in which congestion rises as a power of the volume-capacity ratio. (This is also the form used in the standard Urban Transportation Planning Program package of computer programs provided by the U.S. Department of Transportation for state and local agencies.) Peak-period volume is V/h, where h is the number of hours of peak travel per year; capacity is $c_w W$, as noted earlier. Hence congestion cost is

$$\text{(6-3)} \qquad c(V,W) = c_1 \left(\frac{V/h}{c_w W} \right)^k ,$$

where k is a constant and c_1 is a cost parameter.

Returns to scale and economies of scope are properties of the long-run cost function. Long-run total cost is defined as the smallest possible value that equation 6-1 can take for any W and D. In other words, it is the cost resulting from an optimal investment policy that chooses highway width and durability—call them W^* and D^*—to minimize equation 6-1. One can see from the separate components of that equation that choosing W^* involves trading off increased capital and maintenance costs with lower congestion cost, while choosing D^* involves trading off increased capital cost with lower maintenance cost (exactly as was done in the analysis of optimal durability in earlier chapters). So the entire concept of long-run costs involves a balancing of construction costs, resurfacing costs, and costs to users. Highway engineers often refer to this idea as minimizing the "life-cycle costs" of highways. For convenience, we show this long-run total cost as a sum of four components, which we call congestion, nonpavement capital, pavement capital, and maintenance, respectively:

$$(6\text{-}4)\quad LRTC(V,Q) = V \cdot c(V,W^*) + r \cdot [k_0 + k_1 W^*] + r \cdot k_2 W^* D^* + r \cdot M(Q,W^*,D^*).$$

Once these cost-minimizing values for highway width and durability are determined, one can use differential calculus on equation 6-1 to measure the incremental cost of increasing either V or Q by one unit. This gives precisely the marginal-cost user charges—for congestion and road wear, respectively—discussed in earlier chapters. It is then straightforward to compute revenues from the charges and to compare them with various components of the long-run cost.

We provide three such comparisons, corresponding to what we believe are conceptually useful categories of public budgets. One is the "pavement budget surplus" introduced previously; it is equal to the excess of revenues from road-wear charges over the public costs related to providing durability, which we measure simply as the capital and maintenance cost of the pavement itself (that is, the last two terms in equation 6-4) less that portion of maintenance representing disruption costs to motorists.[9] Analogously, we define the "capacity budget sur-

9. As explained in the source note to table 3-1, our estimate of parameter k_m (to

plus'' as the excess of congestion-charge revenue over the public costs related to providing capacity; we define these public costs as all capital and maintenance costs (that is, the last three terms in equation 6-4) less disruption costs, on the assumption that all capital costs would be avoided if traffic volume were zero. Finally, we define the ''combined budget surplus'' as simply total revenues minus total public costs, where public costs are again all capital and (nondisruption) maintenance costs.

It is only this combined budget that matters for tracing the effects of our policies on real public budgets. The separate pavement and capacity budgets are somewhat arbitrary, but are conceptually useful in enabling us to relate the combined budget surplus to various measures of scale economies that are familiar from other work. Our choices for what to include in the pavement and capacity budgets reflect a view that trucks are in a sense the optional product. More precisely, it is possible to build a road to handle a large traffic volume and an insignificant number of loadings, but not vice versa; hence we view the fixed cost k_0 in highway construction as part of the cost of capacity, not durability. This allocation is consistent both with our analysis of pavement budgets in chapter 3 and with the empirical work on scale economies reviewed in the previous section.

In the analysis of multiproduct industries, four types of returns to scale can be identified and related to each other.[10] The first is product-specific returns to scale for traffic volume, S_V, defined as the average cost of V divided by its marginal cost; here average cost is the difference (for a given Q) between total cost at V and total cost if V were zero, divided by V. Equivalently, S_V is the extra cost occasioned because V is not zero, divided by the revenue from an optimal congestion charge. A value greater than one indicates increasing returns to scale.

Second is product-specific returns to scale for traffic loadings, S_Q, defined as the average cost of Q divided by its marginal cost; here average cost is the difference (for a given V) between total cost at Q and total cost if Q were zero, divided by Q. Equivalently, S_Q is the extra cost occasioned because Q is not zero, divided by the revenue from an optimal road-wear charge. A value greater than one indicates increasing returns

---

which maintenance cost M is directly proportional) includes a simple markup to represent disruption costs to motorists; for urban roads the markup is 20 percent.

10. Bailey and Friedlaender, ''Multiproduct Industries,'' pp. 1030–31.

to scale. Third is economies of scope, S_c, defined as the difference between the costs of providing separately for V and Q and the cost of providing jointly for V and Q, as a fraction of the latter. Equivalently, S_c is the proportion by which total costs would rise if joint production were not possible. A value greater than zero indicates economies of scope. Last is multiproduct returns to scale, S_m, defined as the ratio of total cost of joint production to the combined revenues from optimal user charges. A value greater than one indicates increasing returns.

These quantities bear a relationship with each other that shows how multiproduct returns to scale depend upon the other three:

$$S_m = \frac{wS_V + (1 - w)S_Q}{1 - S_c}, \tag{6-5}$$

where w is the proportion of user charges accounted for by congestion charges:

$$w = \frac{V \cdot MC_V}{V \cdot MC_V + Q \cdot MC_Q}. \tag{6-6}$$

This formulation makes precise what we said earlier: multiproduct scale economies are a weighted average of the product-specific scale economies, but are reduced if there are diseconomies of scope (that is, if S_c is negative).

If we had made different choices about which fixed costs to associate with each output, the measures S_V, S_Q, and S_c would be different; but S_m would remain unchanged and, as is clear from its definition, it alone determines the combined budget deficit.

We calculated W^*, D^*, and the various budgetary and scale-economy measures for several values for traffic volume and traffic loadings, covering a realistic range of conditions, for two sets of parameters, one representing a typical urban expressway of rigid pavement and the other an urban arterial of flexible pavement. Some parameters are identical to ones used earlier (see table 3-1), and the others are listed in table 6-1.

Results

Tables 6-2 and 6-3 show the financial implications of our model when applied to an urban expressway and a principal urban arterial, respectively. The traffic volumes and loadings for which we calculated results

Table 6-1. *Parameters for Model of Multiproduct Returns to Scale*

Parameter	Description	Value
m	Annual rate of increase in pavement roughness due to aging[a]	0
λ	Proportion of loadings occurring in outer lane[b]	0.7
k	Power law for congestion cost as function of volume-capacity ratio[c]	
	Urban expressway	4.0
	Urban arterial	2.5
c_1	Congestion cost when volume-capacity ratio is one (dollars per vehicle-mile)[d]	0.125
k_0	Fixed capital cost (dollars per mile)[e]	
	Urban expressway	149,651
	Urban arterial	93,183
k_1	Nonpavement capital cost that varies with width (dollars per lane-mile)[e]	
	Urban expressway	1,468,462
	Urban arterial	871,599
h	Number of peak hours each direction per year[f]	500
c_w	Capacity of a standard lane (passenger-car-equivalents per hour)[g]	
	Urban expressway	1,800
	Urban arterial	1,080

a. In chapter 3 we used 0.04 for flexible pavements; here we use zero to simplify the calculations.

b. This is the value for expressways of six or more lanes; we believe it is reasonable for arterials of that width or less, since trucks more often travel in the left-hand lanes on arterials.

c. William Vickrey has claimed that traffic data fit equation 6-3 in the text with the following values of k: 4.5 for the Lincoln Tunnel; 3 to 4 for city traffic in Los Angeles, New York, and Philadelphia; and 3 to 5 or even higher in "situations where considerable congestion exists." See William S. Vickrey, "Pricing in Urban and Surburban Transport," *American Economic Review*, vol. 53 (May 1963, *Papers and Proceedings, 1962*), p. 462; "Pricing as a Tool in Coordination of Local Transportation," in National Bureau of Economic Research, *Transportation Economics: A Conference of the Universities—National Bureau Committee for Economic Research* (Columbia University Press, 1965), pp. 285–86; and "Congestion Theory and Transport Investment," *American Economic Review*, vol. 59 (May 1969, *Papers and Proceedings, 1968*), pp. 251–52 (quotation is from p. 252). We have found no documentation of these estimates. The value of k used in the Urban Transportation Planning Program (UTPP) is 4.0. Marvin Kraus and coworkers have used, without explanation, a value 2.5: see Marvin Kraus, Herbert Mohring, and Thomas Pinfold, "The Welfare Costs of Nonoptimum Pricing and Investment Policies for Freeway Transportation," *American Economic Review*, vol. 66 (September 1976), p. 544; and Marvin Kraus, "Indivisibilities, Economies of Scale, and Optimal Subsidy Policy for Freeways," *Land Economics*, vol. 57 (February 1981), pp. 116, 118. We expect k to be higher on freeways than on arterials because congestion on arterials occurs more gradually as traffic volume rises; hence our compromise values of 4.0 on freeways, 2.5 on arterials.

d. This value assumes that average travel time rises one minute per mile as volume-capacity ratio rises from zero to one (for example, freeway speed might decrease from 60 mph to 30 mph, and arterial speed from 40 mph to 24 mph); and that time is valued at $7.50 per vehicle-hour.

e. The values for k_0 and k_1 are chosen to make equation 6-2 for capital cost a linear approximation to the estimated equations 9 and 10 in Theodore E. Keeler and Kenneth A. Small, "Optimal Peak-Load Pricing, Investment, and Service Levels on Urban Expressways," *Journal of Political Economy*, vol. 85 (February 1977), pp. 7–9, inflated from 1972 prices to 1982 prices by a factor 2.287 derived from the composite price trend for federal-aid highway construction, FHWA, *Highway Statistics, 1986* (Department of Transportation, 1987), p. 58. The linear approximation is taken around $W = 3$. The result is that, for urban expressways, capital cost per lane-mile for a six-lane freeway is equal to the value of $540,545 given in Keeler and Small, table 2, for "urban-suburban (outside central city) freeway," inflated as just stated and with 32.3 percent added for right of way as indicated in Keeler and Small's equation 10; k_0 is equal to this amount multiplied by $[(1/s) - 1]$, where s is the degree of scale economies, and k_1 is then determined using the previously determined value of k_2 (see chapter 3) and assuming a standard freeway pavement thickness of $D = 10$. For urban arterials, the corresponding values are $331,815 (applying to "conventional arterial streets or roads within city limits"), 34.2 percent, and $D = 5.3$.

f. Assumes two hours a day in each direction, 250 days a year.

g. Expressway capacity under ideal configuration is 2,000 cars an hour, according to Transportation Research Board, *Highway Capacity Manual*, Special Report 209 (Washington, D.C.: National Research Council, 1985), p. 3-4; we have reduced it by 10 percent to account for nonideal conditions such as weaving at exits and entrances, narrow shoulders, and curves. Arterial capacity is given in the same publication (p. 11-11) as 1,600 cars an hour multiplied by the fraction of time that the traffic signals are green; we take that fraction to be 0.75 for a major arterial, and again subtract 10 percent for nonideal conditions.

Table 6-2. *Results for Model of Multiproduct Returns to Scale: Urban Expressway*

Item	Low volume		Medium volume		High volume	
	Low loadings	*High loadings*	*Low loadings*	*High loadings*	*Low loadings*	*High loadings*
Assumptions						
V (millions of peak pces a year)	1.33	1.33	2.00	2.00	2.67	2.67
Q (millions of esals a year)	0.25	1.00	0.25	1.00	0.25	1.00
Results						
*W** (lanes)	2.02	2.01	3.03	3.01	4.04	4.02
*D** (inches)	9.95	13.10	9.95	13.10	9.95	13.10
Peak volume-capacity ratio	0.73	0.74	0.73	0.74	0.73	0.74
User charges						
Cents a peak pce-mile	14.5	14.8	14.5	14.8	14.5	14.8
Cents an esal-mile	1.2	0.4	1.9	0.6	2.5	0.8
Revenues (thousands of dollars a year per lane-mile)						
Congestion charges	193.0	196.9	289.5	295.3	386.0	393.8
Road-wear charges	3.1	4.0	4.6	5.9	6.2	7.9
Total	196.1	200.8	294.1	301.3	392.2	401.7
Components of long-run total cost (thousands of dollars a year per lane-mile)						
Congestion	48.3	49.2	72.4	73.8	96.5	98.4
Nonpavement capital	186.9	186.0	275.9	274.6	364.9	363.1
Pavement capital	14.1	18.5	21.2	27.8	28.3	37.0
Maintenance	0.9	1.3	1.4	2.0	1.9	2.7
Total	250.2	255.1	370.9	378.2	491.5	501.2
Budget surplus (thousands of dollars a year per lane-mile)						
Capacity	−8.8	−8.8	−8.7	−8.6	−8.7	−8.5
Pavement	−11.8	−15.7	−17.7	−23.5	−23.7	−31.3
Combined	−5.7	−4.8	−4.1	−2.7	−2.5	−0.6
Scale economies						
S_V (Product-specific − *V*)	1.037	1.036	1.025	1.024	1.019	1.018
S_Q (Product-specific − *Q*)	4.926	5.072	4.926	5.072	4.926	5.072
S_c (Scope)	−0.061	−0.079	−0.061	−0.080	−0.062	−0.080
S_m (Multiproduct)	1.024	1.020	1.012	1.008	1.006	1.002

Source: Authors' calculations.

imply optimal highway widths of two to four lanes in each direction, resulting in peak volume-capacity ratios very close to 0.74. Optimal pavement thicknesses are similar to those shown in earlier chapters. So are user charges: around 15 cents an auto-mile for congestion, and 0.4 to 4.0 cents a standard-axle-mile for road wear.

Table 6-3. *Results for Model of Multiproduct Returns to Scale: Urban Arterial*

Item	*Low volume*		*Medium volume*		*High volume*	
	Low loadings	*High loadings*	*Low loadings*	*High loadings*	*Low loadings*	*High loadings*
Assumptions						
V (millions of peak pces a year)	0.40	0.40	0.80	0.80	1.60	1.60
Q (millions of esals a year)	0.10	0.50	0.10	0.50	0.10	0.50
Results						
W^* (lanes)	1.00	0.99	2.00	1.98	4.00	3.96
D^* (structural number)	4.97	6.27	4.97	6.27	4.97	6.27
Peak volume-capacity ratio	0.74	0.75	0.74	0.75	0.74	0.75
User charges						
Cents a peak pce-mile	14.8	15.1	14.8	15.1	14.8	15.1
Cents an esal-mile	1.0	0.2	2.0	0.5	4.0	1.0
Revenues (thousands of dollars a year per lane-mile)						
Congestion charges	59.1	60.4	118.2	120.7	236.3	241.4
Road-wear charges	1.0	1.2	2.0	2.4	4.0	4.9
Total	60.1	61.6	120.2	123.1	240.3	246.3
Components of long-run total cost (thousands of dollars a year per lane-mile)						
Congestion	23.6	24.1	47.3	48.3	94.5	96.6
Nonpavement capital	57.9	57.4	110.1	109.3	214.7	212.9
Pavement capital	6.5	8.1	13.0	16.3	26.0	32.6
Maintenance	0.3	0.4	0.6	0.8	1.2	1.5
Total	88.3	90.1	171.0	174.6	336.4	343.6
Budget surplus (thousands of dollars a year per lane-mile)						
Capacity	−5.5	−5.5	−5.5	−5.5	−5.4	−5.3
Pavement	−5.7	−7.2	−11.5	−14.5	−23.0	−29.0
Combined	−4.5	−4.3	−3.5	−3.0	−1.4	−0.5
Scale economies						
S_V (Product-specific − V)	1.068	1.066	1.034	1.033	1.017	1.017
S_Q (Product-specific − Q)	6.867	7.154	6.867	7.154	6.867	7.154
S_c (Scope)	−0.078	−0.097	−0.081	−0.100	−0.082	−0.101
S_m (Multiproduct)	1.055	1.051	1.021	1.018	1.005	1.002

Source: Authors' calculations.

We would like to compare these volume-capacity ratios with those found on current highways, but we have insufficient data on the wide range of cost and demand conditions met in actual practice. For example, we know from the fundamental law of highway congestion, discussed in the previous chapter, that very complex demand patterns would have to be understood before such a comparison could tell us what the optimal capacity of existing roads might be.

The revenue figures in tables 6-2 and 6-3 show clearly that congestion charges would bring in far more revenue than road-wear charges—typically fifty or sixty times as much. This difference reflects the much greater cost, as well as smaller scale economies, in providing capacity compared with providing durability.

The pavement deficit is proportionally very large: about 80 percent of public durability costs on urban expressways and 85 percent on urban arterials. These numbers are consistent with the finding of chapter 3 that optimal pricing and investment would lead to revenues covering well under half the pavement costs.

The capacity budget has a much smaller deficit as a proportion of public capacity costs—2–4 percent for expressways, 2–9 percent for arterials. But because those capacity costs are so large, the absolute capacity deficit is still substantial. This deficit is highly dependent upon the assumed degree of returns to scale in construction.

As suggested earlier, however, the integration of the two user charges reduces the deficits considerably. The deficit in the combined budget for a highway serving both light and heavy vehicles is, in almost every case, less than the deficit for either taken separately. At medium volume and high loadings, the combined deficit is only one-third the pavement deficit for the expressway, and just over half of it for the arterial. This finding verifies the point made earlier that diseconomies of scope result in some double-charging for pavement costs, which helps compensate for the deficits in the separate pavement and capacity budgets. The remaining deficit is generally not more than a few percent of total costs, and could easily be closed by retaining small registration fees or fuel taxes.

The diseconomies of scope themselves are on the order of 6–10 percent—that is, modest but significant. As a result, multiproduct economies of scale are close to one throughout, ranging between 1.00 to 1.06, and are always below the product-specific economies of scale in traffic volume.

The existence of diseconomies of scope suggests the possible desirability of producing the two products separately. Of course, our two "products," traffic volume and traffic loadings, do not correspond precisely to cars and trucks, and there is no way to produce loadings without some volume. Nevertheless, one could accomplish the same end by restricting some high-capacity roads to cars and using a sparser network of routes to handle trucks; this configuration of roads might save substantial costs provided the resulting circuity of truck travel were not too great.

Three further arguments, not accounted for by our model, support the case for more separation of car and truck traffic. First, much of the expense of a general-purpose highway is for making bridges, underpasses, shoulders, and other structures suitable for larger and heavier vehicles. It has been estimated that 23 percent of the cost of building an urban expressway would be saved if the road were restricted to autos.[11] Second, traffic safety would probably be improved, because speed differentials between cars and trucks appear to contribute substantially to accidents.[12] Third, traffic flow on congested roads is improved by eliminating vehicles with slow acceleration—an advantage cited in reserving parts of the New Jersey Turnpike for cars.[13]

Needless to say, the precise results of our model depend on the parameters we have chosen. As our discussion indicates, one of the most crucial is the degree of scale economies with respect to width in highway construction. To show explicitly how our results depend on this parameter, we have performed calculations under the two alternative assumptions mentioned earlier—constant returns, and increasing returns as estimated by Kraus—which bracket the value of 1.03 assumed thus far. Table 6-4 shows what happens to some key results when traffic is set at the medium volume and high loadings and this scale-economy parameter is varied. (The middle column in table 6-4 repeats information from the fourth column of tables 6-2 and 6-3). Under constant returns in capacity, the diseconomies of scope more than offset the increasing returns in durability, leading to a surplus in the combined budget. Under the more strongly increasing returns in capacity shown in the last column, however, the shortfall in congestion-charge revenues dominates the calculations, leaving a combined budget deficit between 13 percent and 20 percent of capital and maintenance costs. Hence if more careful investigation should show that Kraus's estimates are correct, more revenue

11. Keeler and Small, "Optimal Peak-Load Pricing," p. 8, calculated from data in John R. Meyer, John F. Kain, and Martin Wohl, *The Urban Transportation Problem* (Harvard University Press, 1965), pp. 204–06.

12. Charles A. Lave, "Speeding, Coordination, and the 55 MPH Limit," *American Economic Review,* vol. 75 (December 1985), pp. 1159–64.

13. Paul M. Weckesser and Kenneth W. Dodge, "Efficient Use of a Busy Roadway," *Traffic Engineering,* vol. 46 (April 1976), pp. 20–24.

Table 6-4. *Budgetary Measures and Multiproduct Scale Economies under Different Assumptions about Returns to Scale in Construction of Highway Capacity*

Item	*Degree of scale economies in capacity*		
	1.00	*1.03*	*1.19*
Urban expressway			
Revenue (thousands of dollars a year per lane-mile)	308.4	301.3	270.5
Capital and maintenance costs (thousands of dollars a year per lane-mile)	302.6	304.4	311.3
Budget surplus (thousands of dollars a year per lane-mile)			
Capacity	0.3	−8.6	−46.7
Pavement	−23.4	−23.5	−24.2
Combined	6.2	−2.7	−40.6
Multiproduct scale economies (S_m)	0.985	1.008	1.122
Urban arterial			
Revenue (thousands of dollars a year per lane-mile)	125.7	123.1	111.8
Capital and maintenance costs (thousands of dollars a year per lane-mile)	123.3	126.4	138.6
Budget surplus (thousands of dollars a year per lane-mile)			
Capacity	0.1	−5.5	−29.1
Pavement	−14.4	−14.5	−15.1
Combined	2.5	−3.0	−26.6
Multiproduct scale economies (S_m)	0.986	1.018	1.172

Source: Authors' calculations. For expressway, assumes traffic volume V = 2.00 million peak passenger car equivalents a year and traffic loadings of Q = 1.00 million equivalent standard axle loads a year. For arterials, assumes V = 0.80 million and Q = 0.50 million.

from local property and sales taxes, vehicle registration fees, or fuel taxes will be required to balance the road budget, but the amount of such revenue required will still be far lower than at present.

Conclusion

Because most highways are built to accommodate both peak traffic volumes and numerous traffic loadings from heavy vehicles, the cost of providing the pavement itself cannot be allocated solely to one or the other. Instead, marginal-cost pricing incorporates some of this cost into both the congestion charge, related to the expense of widening the road, and the road-wear charge, related to the expense of increasing its durability. This double charging is a reflection of diseconomies of scope, which are modeled in this chapter to determine the costs and revenues that would pertain to typical urban highways under optimal pricing and investment policies.

The shortfall in revenues predicted from viewing congestion pricing and road-wear pricing separately becomes considerably smaller when they are viewed together. Our model suggests that for major urban roads with typical traffic, revenues from marginal-cost user charges would cover at least 80 percent of long-term capital and maintenance costs, making possible substantial cuts in such forms of road taxation as license fees and fuel taxes.

The finding of diseconomies of scope raises the question of whether to separate auto and truck traffic. Although our model cannot answer this question directly, our findings are provocative enough to warrant a closer look at the advantages and disadvantages of such a strategy, which could save considerably on construction and maintenance costs of high-volume roads. Accounting for the safety and operational advantages of separating heavy and light vehicles would further strengthen the case for doing so.

Congestion is far more important than road wear in determining aggregate costs and revenues from the policies we propose. Pavement costs, including construction and periodic resurfacing of the pavement itself, are less than 15 percent of the road authority's cost in our simulations, and road-wear charges account for only about 2 percent of the combined revenues from the two pricing policies. This finding underscores both the enormous importance of congestion in determining highway expenditure policy and the somewhat misplaced emphasis on heavy vehicles in most cost-allocation studies. The solution, however, is not to increase existing taxes on cars relative to those on trucks; it is to impose congestion charges where they are needed.

CHAPTER SEVEN

A New Highway Policy

FACED WITH a congested and physically deteriorating road system, the United States can no longer rely on current highway policy to finance and manage its roads. In place of the current policy we propose one built on two economic principles: efficient pricing to regulate demand for highway services and efficient investment to minimize the total public and private cost of providing them. The close relationship of the two provides a way to analyze them as an integrated financial package.

Road-Wear Charges and Investment in Durability

The first set of user charges that we recommend—road-wear charges—applies the principle of charging vehicles for the wear and tear they inflict on pavements based on the damaging power of their axle loads. The current user charges on heavy vehicles, based mainly on weight, cannot approximate this goal because pavement damage depends critically on axle loads, not on total vehicle weight.

Implementing road-wear charges is feasible and not particularly expensive. New Zealand has operated such a system for years.[1] Enforcement in the United States should be no more costly than it is for the weight-distance taxes currently in place in ten states. These taxes already require record keeping by firms, record auditing by the tax authority, and occasional weight checks by the highway patrol. Indeed, a recent federal study shows that administrative and compliance costs for a national tax based on axle weight and distance would be little more per vehicle than for the current federal heavy-vehicle use tax, since mileage

1. New Zealand Ministry of Transport, *Road User Charges* (Auckland: Government Printing Office, 1988). According to the Auckland Regional Authority's Director of Transport, New Zealand's axle-based system is based on the fourth-power law of pavement damage and was instrumental in persuading the Auckland Regional Authority to add a third axle to its buses (private communication).

records are already kept by the carriers.[2] (Our tax applies to more vehicles, but even that would raise costs no more than $15 million a year, a negligible amount compared with the benefits we estimate.)[3] Recent technological advances in truck weighing and monitoring, made possible by new devices for weighing trucks in motion and by improvements in microelectronic identification, should make enforcement even easier.[4]

While current road taxes actually encourage the use of trucks that severely damage the roads, our proposed charges would induce firms to use less damaging vehicles, load existing vehicles less heavily, and, perhaps, alter the logistics of their entire operation to use roads most suited to heavy trucks. Our estimates of the benefits of road-wear charges are based just on the first of these responses, for private and common-carrier trucking only. Even so, we find (table 4-1) that revised road-wear charges for existing pavement could reduce traffic loadings 48 percent, for a long-term saving of $6.4 billion a year in resurfacing costs at 1982 prices.

To be fully effective, the charges should apply to all heavy vehicles—public or private, trucks or buses or construction equipment. Publicly owned vehicles are often exempted from road taxes on the grounds that charging them just shifts money from one government account to another, but in this case exempting those vehicles squanders the incen-

2. U.S. Department of Transportation, *The Feasibility of a National Weight-Distance Tax* (December 1988), pp. IV-19, V-15, and V-16.

3. From DOT, *Feasibility,* p. IV-19, annual administrative costs might rise from $11.5 million to $31.3 million; and from pp. II-26 and V-28, the "high estimate" of annual compliance cost is $20.4 million a year, compared with $24.9 million for the current heavy-vehicle use tax, truck excise tax, and tire excise tax. We have used here DOT's estimate for compliance cost for a tax based on registered gross vehicle weight, not registered axle weight; the reason is that their estimate for the latter is based on registering every axle, whereas we agree with their discussion on p. V-16 that a simpler implementation based on gross vehicle weight and axle configuration would be adequate and would be no more difficult to comply with than the weight-distance tax.

4. These devices are being tested in a cooperative project involving the federal government and several states. Originally known as the Heavy Vehicle Electronic License Plate (HELP) project, it has now been subsumed into the broader "Crescent Project," which includes a binational demonstration program on a crescent-shaped corridor of continuous interstate routes running from British Columbia through California to Texas. See Loyd Henion and Barbara Koos, "The Heavy Vehicle Electronic License Plate System Development Program: A Progress Report," *Transportation Research Forum Proceedings,* vol. 27, no. 1 (1986), pp. 188–92; and Peter Davies and Fraser K. Sommerville, "Development of Heavy-Vehicle Electronic License Plate Concept," *Transportation Research Record,* no. 1060 (1986), pp. 121–27.

tive to reduce axle loadings. It is better to charge each public agency for the road wear it imposes, let it determine the best way to trade that cost against its other needs, and simultaneously provide a more accurate accounting of the true costs of government programs.

The theoretically ideal road-wear charge would vary by road type, reflecting the much greater vulnerability of thin roads to damage, but would be complicated to administer. Although simplifying the charge radically to a single tax schedule applying to all roads would retain a surprisingly large proportion of the benefits of such a user charge, a system of at least two schedules, for example one schedule for major arterials and one for all other roads, would capture more benefits and still be practical to administer. Our proposed system of road-wear charges would permit some relaxation of current weight regulation, although for safety purposes continued use of weight limits for specific roads and occasionally for specific seasons would still be required.

Our analysis of pavement durability, using standard economic techniques and a new statistical analysis of road-test data, suggests that a substantial increase in durability could be achieved at modest cost and would lower the total costs of building and maintaining pavements over their life cycle. We estimate that even with no change in user charge, an increased capital outlay with annualized value of $2.2 billion would lower annualized resurfacing costs by $8.5 billion, for a net saving of $6.3 billion (table 4-2). The largest increases in durability are needed on heavily traveled urban interstates and other principal arterials. Our analysis does not include bridges.

Not only does increased durability lower costs in the long run, but because more durable roads mean lower user charges, it makes the needed road-wear pricing much more attractive politically. The impact of increased durability and road-wear pricing together on the trucking industry is approximately neutral—in fact, slightly favorable on average—yet the two would improve the public's highway budget balance an estimated $6.3 billion a year (table 3-13) if taxes charged automobiles and pickup trucks were not changed. This dramatic savings occurs because the restructured road-wear charges induce shifts to trucks with more axles, thereby reducing loadings 38 percent. As a result, with both policies in place, additional annualized capital expense is only $1.3 billion, yet the annual saving in life-cycle costs is $8.2 billion (table 3-7).

With the durability improvements that we recommend, the new road user charge for a typical 80,000-pound fully loaded five-axle tractor-semitrailer combination in intercity use, for example, would be less than two-thirds the average fuel and weight-related taxes and registration fees currently paid; whereas the charge for a more damaging 33,000-pound two-axle van in urban use would triple current taxes and fees (tables 3-4 and 3-5). It is precisely such shifts in relative taxes that provide incentives to use less damaging vehicles. As this example suggests, the use of the larger trucks would increase, not decrease. In fact, we estimate that use of every truck type except two-axle single-unit trucks would increase slightly.[5] The SU2, both the most numerous and the most damaging per ton of cargo, would decline 5 percent in intercity operations and 7 percent in urban operations (table 3-9).

There is a legitimate safety concern about the large double trailers whose use might be encouraged by this policy. At present, there is no conclusive evidence that double trailers compromise safety. Studies claiming to have found such evidence have often failed to control for weather, driver's record, safety practices of the firm, mechanical condition and age of the truck, size of load relative to truck capacity, and stringency of safety inspections.[6] We recommend not only continued research into safety, but direct regulation where problems are discov-

5. This runs directly contrary to the finding of the latest Highway Cost Allocation Study, which claims that single-unit trucks are the one class of trucks significantly overcharged at present. See FHWA, *Final Report on the Federal Highway Cost Allocation Study* (Department of Transportation, 1982), pp. I-10, I-13; Roger Mingo and Anthony R. Kane, "Alternative Equity-Based Methods for Allocating Highway Costs among Users," *Transportation Research Forum Proceedings,* vol. 23 (1982), pp. 306–12, especially p. 309; Alice M. Rivlin, testimony before the U.S. Congress, Senate Committee on Environment and Public Works, August 18, 1982; and U.S. Congressional Budget Office, *New Directions for the Nation's Public Works* (GPO, September 1988), p. 9.

6. For example, Howard S. Stein and Ian S. Jones, "Crash Involvement of Large Trucks by Configuration: A Case-Control Study," *American Journal of Public Health,* vol. 78 (May 1988), pp. 491–98. For other reviews of the safety issue, see Committee for the Twin Trailer Truck Monitoring Study, Transportation Research Board, *Twin Trailer Trucks: Effects on Highways and Highway Safety,* Special Report 211 (Washington, D.C.: National Research Council, 1986), which argues that use of double trailers would not increase accidents; Robert E. Skinner and Joseph R. Morris, "Monitoring the Effects of Double-Trailer Trucks," *Transportation Research News,* no. 114 (September–October 1984), pp. 15–21; FHWA, *The Feasibility of a Nationwide Network for Longer Combination Vehicles,* Report of the Secretary of Transportation to the U.S. Congress (Department of Transportation, May 1985); and Organization for Economic Cooperation and Development, *Impacts of Heavy Freight Vehicles* (Paris: OECD, 1983).

ered. Additional safety regulation required because of a switch to larger trucks would affect our proposed policy little because most benefits can be realized by shifting to same-sized vehicles with more axles.

Congestion Charges and Investment in Capacity

The costs of laying and maintaining pavements is dwarfed by the cost of providing sufficient capacity to meet peak-hour demand for travel. Economists have long understood that such demand could be moderated by charges specific to those times and places where congestion is severe. Although our estimates of the possible revenues and welfare improvements from congestion pricing are much less precise than those for road-wear charges, it is clear that the potential saving is enormous, with annual revenues of tens of billions of dollars and annual time savings worth perhaps $5 billion a year, even with no changes in capital spending.

Recent technological advances make fine-tuned peak-period user charges feasible and should erase forever the image of thousands of toll booths standing sentinel over the urban landscape. Systems as varied as windshield stickers and electronic vehicle identification have been field-tested on a variety of U.S. toll facilities and on a citywide scale in Singapore and Hong Kong. There seems little doubt that a reliable and unintrusive charging system is now possible.

We also believe that the political climate is more favorable to congestion pricing than ever before.[7] The sheer magnitude of congestion and its resistance to attack by other policies are creating greater public willingness to try previously unacceptable measures. Furthermore, a growing concern with highway finance makes pricing measures a natural place to look for innovations. We have attempted to bolster the case by demonstrating how a combined package of road-wear and congestion charges can come closer to balancing the highway budget than either could alone. In fact, such a package could largely replace traditional

7. Some might argue that the failed attempt to institute congestion pricing of runways at Logan (Boston) Airport suggests otherwise. However, as Steven A. Morrison and Clifford Winston point out in "Enhancing the Performance of the Deregulated Air Transportation System," *Brookings Papers on Economic Activity, Microeconomics, 1989,* the revised prices at Logan were applied only to general aviation and were not differentiated by time of day. Correct, and politically sensitive, application of congestion pricing requires that all users be charged appropriately and that prices be differentiated by time of day.

revenue sources such as fuel taxes and license fees, while freeing general local revenues for other purposes. Hence congestion charges can simultaneously overcome the current gap in highway finance, replace many current highway charges, improve the financial status of local governments, and cover the "pavement deficit" that would result from efficient road-wear charges on a highway system of optimal durability.

Congestion charges would supplement and in some instances replace the complex traffic-management measures increasingly proposed to cope with traffic congestion. For example, by raising the private cost of each vehicle-mile driven on congested roads, they would provide strong incentives for workers to carpool, take transit, or live closer to jobs, eliminating the need for high-occupancy special lanes. They would also provide market signals of the demand for new investment, which is one of the main objectives of recent suggestions for more privatization of highways.[8]

We have not attempted to analyze the road capacities that would characterize an efficient road network with congestion pricing. To do so would require detailed knowledge of the traffic volumes by time of day on numerous road segments, the costs and feasibility of possible widening measures, and the costs of alternative designs for planned new roads, as well as a better knowledge of returns to scale in highway construction, including some of the factors noted in chapter 6. Data collected as part of the Federal Highway Administration's Highway Performance Monitoring System could now permit a start on such a project, which would provide important information about the savings in new construction that might be permitted by a policy of widespread congestion pricing.[9]

However, some preliminary speculation about optimal road capacities is possible even now. One of the primary effects of congestion pricing is to reduce peak-period traffic, which determines the needed highway

8. For some of the arguments in favor of private ownership of highways, see Steven A. Steckler, "Privatization of Highways and Bridges," in Privatization Task Force, *Federal Privatization: Toward Resolving the Deficit Crisis* (Washington, D.C.: Privatization Task Force, June 1988); and a variety of articles in *Private-Sector Involvement and Toll Road Financing in the Provision of Highways, Transportation Research Record,* no. 1107 (1987). In our view, publicly provided and efficiently priced highways are a better solution than franchised monopolies that tend either to set inefficiently high prices or to sink under the weight of oppressively detailed regulation.

9. For an example of how such data can be used to estimate aggregate effects of measures to reduce congestion, see Jeffrey A. Lindley, "Urban Freeway Congestion: Quantification of the Problem and Effectiveness of Potential Solutions," *ITE Journal,* vol. 57 (January 1987), pp. 27–32.

capacity. The empirical studies of congestion pricing described in chapter 5 suggest that plausible congestion tolls would reduce peak traffic volumes 10 percent to 25 percent on many congested highways. Applied to existing roads, the projected reduction could tip the balance so as to make many widening projects unnecessary; applied to new roads, it would make possible smaller and cheaper facilities in many cases.

A road that permits trucks must have lane widths, clearance heights, bridge strength, pavement, and other features capable of handling these heavier and larger vehicles. The best road network may thus contain a denser set of routes for autos than for trucks: in other words, it may contain many auto-only roads. This issue deserves careful scrutiny, especially in light of our recommendation that roads built for trucks be made more durable than they are now.

Managing the Transition

Detailed year-by-year analysis would be required to describe fully the budgetary implications of moving from current policy to a new regime. We believe that the Highway Performance Monitoring System could be adapted to do so, but such a project is beyond our scope. We content ourselves with some observations about the broad outlines of the transition.

First, with federal user charges scheduled to expire in 1993, the time is right to change to axle-load-based user charges for heavy vehicles. Congress has already expressed interest in the concept, and its feasibility has been extensively studied.[10] In theory, the charges should at first be set at current short-run marginal costs, which for most heavy vehicles would be much higher than current user charges (tables 3-4 and 3-5); over the years, as road durability is increased through further investment, the charges would be lowered. In practice, it may make more sense to start with axle-based charges midway between current and ideal charges and to accompany the new system with a definite schedule for reducing the charges to reflect planned improvements to road durability.

Road durability cannot be increased in one grand attack. There are two steps. The first is an immediate reassessment of durability standards for all new highways. Current design guides should be reexamined in

10. DOT, *Feasibility*.

light of our findings that the design equations estimated in the early 1960s are inaccurate and that optimal pavement lives are longer than the planning horizons often adopted.

The second step is to upgrade existing highways. The best way to do that is probably to increase thickness, beyond what would normally be added to restore original strength, each time a pavement is resurfaced. We have not explicitly analyzed the cost, but it seems to be about the same as the incremental pavement cost at new construction that was the basis for our estimates in chapter 2.[11]

Congestion is mainly a local matter, and we believe that congestion pricing can be successfully introduced only through local political processes. The federal government, however, has an obligation to encourage the most economical use of its funds. To this end, we recommend that all federal grants for capacity-increasing highway improvements require a traffic management plan that includes full consideration of congestion pricing, both on the road for which funds are proposed and on competing roads, along with an analysis of how much capital needs might thereby be reduced. One precedent for such a policy is the innovation in hospital siting that ties funds for new facilities to state standards aimed at more efficient use of existing hospitals.[12] Such a policy is also consistent with recent calls for "performance-based decisions" as a general strategy for improving public works management.[13]

The federal government can also help generate a more solid research base for predicting the results of congestion pricing. As already noted, the Highway Performance Monitoring System includes data that could be used to estimate aggregate effects. On a more case-by-case level, the most pressing research need is for better models of the demand response to congestion pricing, especially through such means as shifts in the time

11. American Association of State Highway and Transportation Officials, *AASHTO Guide for Design of Pavement Structures* (Washington, D.C.: AASHTO, 1986), section 5.3; Michael J. Markow and Wayne S. Balta, "Optimal Rehabilitation Frequencies for Highway Pavements," *Transportation Research Record*, no. 1035 (1985), pp. 31–43, especially pp. 35–36; Michael S. Mamlouk and Boutros E. Sebaaly, "Overlay Thickness Design for Flexible Pavement," *Transportation Research Record*, no. 993 (1984), pp. 63–67.

12. Louise B. Russell, *Technology in Hospitals: Medical Advances and Their Diffusion* (Brookings, 1979).

13. National Council on Public Works Improvement, *Fragile Foundations: A Report on America's Public Works* (GPO, February 1988), pp. 114–20.

of day of travel and, over a longer period, in residential and workplace locations.

A city or county wishing to adopt congestion pricing should provide explicit plans for using the revenues, preferably in cooperation with its state government. Such a strategy will increase the political acceptability of congestion pricing. Our analysis in chapter 6 suggests that congestion pricing revenues could substitute for most or all of the existing registration fees, fuel taxes, sales taxes, and property taxes used for urban highways. Particularly important, we believe, is eliminating the drain that highways now place upon local general revenues. We suggest that congestion-pricing proposals include an explicit accounting of the resulting improvement to the balance in local general budgets and of exactly what the improvement will "buy"—for example, lower local taxes or improved specific local services. There would be revenue left over to cut state registration fees and fuel taxes, although doing so would require solving distributional problems between urban and rural areas.

At the national level, the decision whether to eliminate the federal fuel tax will be complicated by competing goals for tax policy. Two of the most prominent are the need for an easily administered source of revenue to reduce the federal deficit and the desire to foster secure energy supplies. More broadly based taxes such as an excise tax on crude oil might meet these goals more effectively, but that is beyond our scope. The lesson to be drawn from our analysis is that for purposes of transportation policy, other user charges far superior to registration fees and fuel taxes can provide most of the needed revenues.

Conclusion

The recent decline of America's road system has been seen as a threat to the nation's quality of life and as yet one more way the United States is starting to fall behind other countries. It is widely believed that the solution calls for hundreds of billions of dollars of capital investment. Such high levels of spending, however, are not only unnecessary; they may lead to a road system that is even more costly to maintain and more of a drain on the public purse than the current system. The pricing and investment policy that we propose instead could eliminate the financial burden now placed upon state and local government general revenues while providing adequate maintenance and continued low-cost shipment of goods by road. It meets the main goals of important political interests and can be implemented with practical, tested procedures and technologies. The nation's road system is too important, and its financial implications too large, for such a fundamental policy change to be further postponed.

Index